WAR AND THE BREED

WAR AND THE BREED

The Relation of War to the Downfall of Nations

by
David Starr Jordan
Chancellor of Leland Stanford Junior University

Illustrated and Abridged with
Foreword by J. W. Jamieson

The Cliveden Press
Washington, D.C.

The Cliveden Press
1133 13th Street, N.W., C-2
Washington, D.C. 20005

ISBN 9-41694-33-X

Library of Congress Cataloging in Publication Data

Jordan, David Starr, 1851-1931.
War and the Breed.

Abridged reprint of original published:
Boston: Beacon Press, 1915.
1. War. 2. Eugenics. 3. Degeneration.
I. Title.
JX1952.J72 1981 303.6'6 81-6075
AACR2

Manufactured in the United States of America

Τοὺς εὐγενεῖς γὰδ κ'αγαΘούς, Φιλεῖ Α'ρης ε'ναίρειν.
(Sophocles: The Phrygians)

Since individuals pass away, parenthood is the supreme factor in the destiny of nations.
(Saleeby)

CONTENTS

FOREWORD

In accordance with Spartan custom, every one of the small body of elite warriors who gave their lives to defend Greece from the Persians at Thermopylae was a volunteer. But, more important, every one was also already the father of at least one son. In this way the ancient Spartan people ensured that their stock should not be enfeebled by the disastrous scythe of war, which always seems to cut deeper into the flower of a nation's manhood while passing more lightly over the poorer stock. In this way, the ancient Spartans, renowned for their courage and patriotic self-discipline, ensured as best they could the survival of their noblest families, which for generation after generation sent forth heroes who gave their lives so that their nation and kinfolk might survive.

War and the Breed, here reprinted in shortened and abridged form, sharply portrays the dysgenic effect of war – and especially of modern war. The reader will be particularly impressed by the prophetic insight of its author, David Starr Jordan, then Chancellor of Leland Stanford University. Jordan wrote *War and the Breed* during the first year of the most disastrous and inexplicable holocaust the Western World has ever known. Completed in March 1915, *War and the Breed* already forecast the critically selective genetic destruction which was to be the prime legacy of World War I. Most of the massive, heroic and totally senseless battles of that fatal war, the internecine slaughter of which was of a magnitude never before equalled in the history of Europe, were yet to be fought. Yet David Starr Jordan foresaw in the winter of 1914-15 the long-drawn Ragnorak of 1915-1918. Half a million men drawn from the flower of Europe had already been killed or maimed in the battle of Ypres, but their number was but an omen, a harbinger of the 10,000,000 strong selective race destruction that was wreaked so permanently upon the greatest nations of Europe in the ensuing years. The Marne, Somme, Ypres, Passchendael, Cambrai, and Verdun, the name of these battles stand as an epitaph to Europe. Those who died so gallantly and so pointlessly in these terrible battles were mostly young men, the fittest that Europe had to offer, very few of whom left any children.

As Jordan emphasizes, Europe was already weakened by a long history of dysgenic war and genetically debilitating colonial enterprises. The origins of World War I have never been clearly traced. However, those who have researched the tragic events recurringly indicate the activities of international revolutionary elements which for decades had sought to overthrow the "old order" in Europe. Following their repeated failure to achieve their goals by direct revolution, it appears to have become evident to them that their best chance lay in a total conflagration which would exhaust the leadership of European society and decimate the patriotic elements which were the most stalwart of their opponents, while passing over the revolutionary elements and providing the latter with an opportunity to spread dissatisfaction and the seeds of rebellion. Thus, the only victors of World War I were the revolutionary forces which seized power in Russia, Hungary, and for a time in Germany, gained strength in many other European countries.

Whatever the causes of World War I might have been, it remains a fact that this conflagration represented the final tragic act in the drama of a dying continent, mortally wounded by the unparalleled genetic and cultural destruction implicit in modern war. Its sequel, World War II, was a ruthless epilogue, selectively sweeping the surviving elements of European genetic talent into those highly selective categories which include the pilots and crew of modern warplanes, of modern tank and armored units, of modern warships and submarines, who met their death on both sides of the rapidly moving front-line. The Battle of Britain, the Desert War, the Normandy invasion, the Battle of the Vosges, the Katyn Forest massacre, the massive battles on the Eastern Front, and the remorseless naval warfare again drained the best blood of the nations of Europe. And again, the only victors were those who did not regard the blood of Europe as its greatest treasure.

The loss of life among the combatants of World War II may or may not have been proportionately as heavy as in World War I, but it was at least equally as selective, if not more so. The population of contemporary Europe, deprived of the descendants of those who were hand-picked to die in these two disastrous wars, is in no way comparable to the European population of 1914. It looks different and it is different. The economic, social, moral and political decline of Europe has genetic and racial components which were foretold by David

Starr Jordan with terrible accuracy when he penned the last words of this book on the twentieth day of March, in that terrible year nineteen hundred and fifteen. Unlike ancient Sparta, which sent its heroes off to die at Thermopylae in defense of Greece and all Europe, modern Europe sent its young men off to war against each other. It compounded this fratricidal crime by a further fratricide against its own posterity by placing the heaviest burden on those who were youthful and still childless. While Sparta survived to raise the sons of those who had died at Thermopylae – sons who would always revere the heroic memory of their dead fathers and who would be inspired to defend Sparta as their fathers had done before them – modern Europe was left with aging parents, bereaved of their descendants by policies whose sole aim appeared to be the destruction of the very nations of Europe. Truly, as the ancients said, those whom the gods would destroy, they first drive to madness.

J. W. Jamieson

I. INTRODUCTION

This book is a study of the effect of war in its relation to the human race and racial development.

It is written in March, 1915, in the midst of the most ruinous war the world has ever known, a war in which the manhood of the race has been wasted as never before in human history, a war from which every nation concerned will awaken exhausted and humiliated for generations to come, its people less courageous, less wise, and feebler in body and spirit than they were before this terrible and senseless sacrifice. The lesson this book hopes to teach was stated by Charles Darwin in 1871. In the *Descent of Man* he says:

> "In every country in which a standing army is kept up, the fairest young men are taken to the conscript camp or are enlisted. They are thus exposed to early death during war . . . and are prevented from marrying during the prime of life. On the other hand, feebler men with poor constitutions are left at home and consequently have a much better chance of marrying and propagating their kind."

It is apparent that armies demand men above the average in physical efficiency. It is plain that the most energetic and intelligent among these make the best soldiers. It is recognized that those who fight best suffer the most in action, while the demands of battle and camp cut off men in the prime of life from normal parenthood. This leaves the weaker elements of one kind or another to be the fathers of the coming generations. By the law of heredity, like the seed is the harvest, and the future of the race repeats the qualities of those war does not use. This thesis is logically without flaw, but to demonstrate its actual validity through the results of the experiments of nations is a task of the most complex character. For a nation does not miss that which it has not had, and all considerations of value of strains of inheritance are mixed inextricably with results of education, organization, commerce, industrialism, opportunity and emigration, influences which may seem to transform a nation in a manner quite independent of the real capacity of its people. However, to show that this thesis is true in fact, as well as sound in theory, is the purpose of this book.

David Starr Jordan

II. FACTORS IN ORGANIC EVOLUTION

Organic Evolution

The term "Organic Evolution" applies to the orderly changes which are now taking place, or which have in the past taken place in living forms, from generation to generation. In this movement the natural history of humanity, its divergence into species, races and strains, forms an integral part. For man is "part and parcel of nature," governed by the same laws of birth, growth and development as the higher animals, laws shared in their degree by our other "brother organisms, the plants," as well. Life in its endless movement we may perhaps liken to a great river, flowing continuously, dividing at times into smaller streams, purifying itself as it flows along and dropping to the bottom its silt and mud.

So too the stream of life diverges, throwing off races and species as it flows, and purifying itself through the process of "Natural Selection," that is, through the survival of those organisms adapted to their environment, while at the same the weak, the ill-begotten and the unfitting are left without progeny. Thus a race of organisms, man, animal, or plant, comes to fit its environment as a river fits its bed.

The river is made up of simple molecules, alike in structure and unchanging in space or time. In the stream of life, on the contrary, no two individuals are quite the same. Each one is plastic, molded by its environment. Each has its reactions, by which it resists environment. Incessantly, the substance of each organism is being worn away, to be swiftly replenished in endless round. In each appears the miracle of birth and life and death; conception, assimilation, growth, differentiation and dissolution. And in the human race, the most complex of all organisms, we have the phenomenon of life developed to the highest degree.

The process of evolution in man, as in the lower animals and plants, represents a series of relations of cause and effect. Plainly each change from generation to generation must have some efficient cause. The causes are known as the *Factors in Organic Evolution*. There may be many of these factors, acting and interacting. Four of them there are, at least, potent in the life of every animal and plant, and from the operations of which no organism can escape. These four are known as *Variation, Heredity, Selection and Segregation.*

Variation

Variation is an attribute universal in the organic world. By its operation, no two individuals of any species, not even of one household, were ever exactly alike. No man, no animal, no germ-cell even, was ever an exact copy of any other. For, with scanty exceptions, throughout nature every organism has two parents, and this phenomenon of double parentage is the leading agency in the promotion of variation. The hereditary traits of each individual, "unit characters" as they are termed, are derived half from one parent and from that parent's ancestry, half from the other and from his ancestry. Each thus begins life as a mosaic of inherited characteristics, and the finished combination can never be twice the same. Further it must be noted that each individual arises in the beginning from the blending of two germ-cells, the male and the female. Each of these germ-cells when matured and fit for conception, contains but half of the original hereditary material of an ordinary cell. It needs union with another half-cell of the opposite sex to begin its organic development. There are many theories as to the origin of variation, but in any event there is no question as to the fact. And on the foundation of variation, the ramifications of species in the organic world take their rise.

Heredity

Heredity is the element of continuity in life. Reproduction is one of the great cardinal functions of living organisms. Heredity is the force or law or fact by which the new organism is like its parentage. Each organism passes through its cycle of life, each made up of changing atoms, changing centers of energy, changing tissues and changing organs, each at the same time having the power to cast off cells which in proper union with like cells and under proper conditions, will develop organisms essentially like the parents, differing only within the limits of the play of variation. "Like produces like," but never quite alike. Heredity and Variation are twins inseparable.

The more uniform the ancestry, the less variation in the progeny. Hence the crossing of two strains within the same species produces less variation than results from the crossing of different but related species. Species widely different cannot be crossed at all.

Among men, individual differences may be much greater than in the lower forms. Animals mate in their own localized groups, whereas men may range widely. Thus it comes about that each individual man or woman is in some sense a hybrid, the result of a mingling of different strains. The laws that govern cross-matings are very complex and imperfectly understood. And no one can tell from the parentage exactly what the offspring will be. But we may be sure of some things it will not be. The stream of heredity will not rise above its source. We shall not gather grapes from thorns nor figs from thistles. Moreover, in heredity, mediocrity is prepotent over genius, and quality of talent over intensity. And again, a well-marked trait may sometimes lie latent or recessive for a generation or more. Still again virile traits, necessarily recessive in the daughter, may reappear in the grandson.

We cannot here take up in any detail the subject of heredity. Enough to say that its laws are essentially the same throughout all living things, the variations being dependent on differences in life history. With the animal forms nearest man in structure, the details of heredity become more and more similar, or even identical with those of man.

Selection

Variation and *heredity* are names for forces resident within the organism. Heredity is never unchecked by variation. Variation is an integral part of heredity. *Selection* and *segregation* on the other hand represent intrusions of the environment. These shape the course of evolution by determining the individual that shall survive or by limiting the range of its mating.

Selection is the process by which an organism which cannot hold its own in its surroundings is destroyed. Only those survive which can live in the actual conditions and only those which do survive are repeated in their progeny. This process as occurring in nature, "*Natural Selection*," as Darwin called it, leads to the "*Survival of the Fittest*," not necessarily of the largest nor the fairest nor the best, but of those who can maintain themselves in the world as they find it. Natural[1] selection, blind choice in nature, goes on constantly, generation after generation, among all living creatures. Environment constantly challenges the right to live, and the organism that overcomes, be it man or fly or seaweed, constantly meets that challenge.

The continuance of any stock demands persistent victory. Each individual man or woman, animal or plant, represents success in escaping from the vicissitudes of all the ages by overcoming the millions of varied changes of environment. Of all our countless ancestors, "numberless out of the endless ages," not a single one, brute or man, ever died in infancy, an amazing record were it not shared by every living creature on the whole Earth!

Segregation

The fourth great factor of evolution is the negative one of *Segregation*, or *Isolation*. The world on which we live has its mountains and its valleys, its rocks and seas, its land and water barriers, its barriers of climate, rainfall, enemies and food. Every species or kind of organism, man included, must fit the surroundings it has or can secure. Thus races are established, not by direct effect of climate or food, so far as we know, but by varying degrees of adaptation to actual conditions. A race once established, the barriers which prevent crossing with other races tend to keep it permanent. Any race or species of man, as of other animals or plants, is thus in a degree a product of Geography.

Physical obstacles prevent intermingling of different types. Localization permits the development of minor peculiarities which would have been lost in the free interbreeding of migration. And in such separations and migrations most of the differences of species or race among organisms have their origin and their permanence. The beginning of races and of species in most cases is due to some initial variation, large or small, emphasized or maintained in environment, and to the protection of some barrier which shuts off a group from the one from which it has diverged. And thus, through limitations to movement and separations due to natural causes, have arisen most of the differences among men as to race, nation, language or religion.

Reversal of Selection

Reversal of selection is the process by which those organisms best fitted to survive under normal conditions are destroyed, while less well endowed types are thus left to reproduce the species. Similar states may be brought about artificially by a change in the conditions of survival. In human affairs such reversal of selection is the necessary result of war[2], as this book will endeavor to show.

War is, of course, not the only reversal factor in modern civilization. There are numerous other elements of varying importance which tend to reverse the natural processes of selection. Chief among these are the effects of *Emigration* and *Immigration*, both briefly treated in another chapter. The various ramifications of *Charity*, wide and unwise, constitute another important factor. By unwise charity, pauperism is helped to perpetuate itself, feeblemindedness becomes the heritage of future generations, and races of idiots and criminals have been created. By wise charity the lives of many of the physically weak have been saved or enriched to the advantage of human character, but not to the physical gain of the race.

The development of a priesthood bound by *Celibacy* has been a loss to religion, debarring piety as it has from reproducing itself in progeny. The conditions of personal success in professional life now tend also to enforce Celibacy on women of superior ability, and have delayed marriage on the part of men. These factors on the whole may work to the advantage of present day society, but certainly at the expense of future racial welfare.

Vice, alcoholism, the use of drugs, the losses of industrialism, are elements to be investigated by the student of social movements, but for the present we have only to consider those factors in reversal of selection which are connected with the main agency, War.

(1) For the interposition of the hand of man modifying conditions of survival, the term Artificial Selection is used. This is discussed further on.

(2) "La guerre a produit en tout temps une selection à rebours." (Novicow.)

III HUMAN HEREDITY

Nature and Nurture

Genetics is the science of birth, development and heredity among living organism. *Eugenics*[1] is the science and art which treats of conditions under which a human being may be well born. As defined by Francis Galton who first devised the word, it had a broader scope than this, including *Euthenics*, or the science and art of being well brought up. Galton's last definition (1904) was the following: 'Eugenics is the science which deals with all influences which improve the inborn characters of a race, also with those which develop them to the utmost advantage.' Recent writers have separated the second element under the name Euthenics. The opposite of *Eugenics*, that is, the promotion of ill-birth, is termed *Dysgenics*. In like fashion, bad rearing is called *Dysthenics*. Eugenics and Euthenics correspond to "*Nature* and *Nurture*," as earlier defined by Mr. Galton. Eugenics and Euthenics must supplement each other. Nature depends on Nurture for normal development. Nurture can modify but not create.[2] With the function of conception, "the gate of gifts is closed." The organism may then make the most of the potentialities heredity has granted it. It can secure nothing more, though use and disuse may greatly modify the relative relations of its inborn powers. Nurture can do wonders with man, but it cannot alter the Nature he is to transmit to his descendants. Bad surroundings may spoil a child of good stock, but good surroundings can never change a bad breed into a good one.

In this fact of the persistence of qualities in the germ plasm lies the hope of the children of the very poor and the very rich. They inherit the possibilities, not the actualities of their parents, spoiled or injured by bad nurture. Their inheritance is a result of what their father and mother might have been, not (disease or mutation excepted) of what they actually are.

In each race group, no matter how small, some families or family strains will be gifted above others. The best of every group constitute the basis of its eugenic progress. There are very many types of fitness, physical, mental, ethical, as many as there are forms of success or of usefulness in life. There are thus numerous elements involved in racial advance. And the best in one of the races considered lower may have greater potential value than the less desirable of a race admittedly higher as a whole.

Meaning of Human Progress

The word "progress" is commonly used with a double meaning, including nurture as well as nature, the advance of education as well as race development. The first of these meanings is entirely distinct from the other. Race improvement is very slow and often thwarted by crimes and blunders. The results of education may be immediate and impressive. But the results of nurture are permanent only as imposed on the solid basis of nature. By training we may increase the force of the individual man. Education gives him access to the accumulated stores of wisdom built up from the experience of the ages. The trained man is placed in a class relatively higher than the one to which he would belong on the score of heredity alone. Heredity carries with it possibilities for effectiveness. Training makes these possibilities actual. Civilization has been defined as "the sum total of those agencies and conditions by which a race may advance independently of heredity." But while education may greatly modify the activities of individuals, and through them those of the nation, these influences are spent on the individual and the social system of which he is a part. So far as science knows, education and training have no part in heredity. The elements in the germ plasm are ancient and persistent, not affected by the vicissitudes of the individual life which bears them from generation to generation. The change in hereditary traits, which is the essence of race progress, as distinguished from progress in civilization, finds its main if not its sole cause in selection.

"Blood Will Tell"

A common expression of the law of human heredity is that "Blood will tell." This means that ancestral traits persistently reappear.

The word "blood" in this sense is figurative only, an expression of the facts of heredity. Some traits, as the phrase goes, "run in the blood." "Blood," says Mephistopheles, "is quite a peculiar juice." And so it is, but not in the degree formerly believed. It was long imagined that blood was the actual physical vehicle of heredity, that the traits of family and race ran literally in the blood itself. This is not the case. Actual blood plays no part in heredity, the transfusion of blood means little more than the transference of food, and the physical basis of the phenomena of inheritance is found in the structure of the germ cell and its contained germ plasm.

But the old word well serves our purpose. Blood which is "thicker than water" is the symbol of race unity. In this sense lies the apparent paradox that blood determines history and history determines blood. For example, wherever Englishmen go they make, whether well or ill, English history. Chinamen make Chinese history. We may note in passing, however, that the climate in which a given tribe may live may affect the activities of its members as individuals or as an aggregate; education may intensify their powers or mellow their prejudices; oppression may make them servile, or dominion make them arbitrary; but these traits and their resultants due to external influences do not "run in the blood," they are not "bred in the bone." Older, deeper set, more permanent than climate or training or experience are the traits of heredity and in the long run it is always "blood that tells."

But even hereditary traits are not immutable. War and conquest, with other selective influences, may modify even these. It is the man who is left who determines the future trend of history. His inborn qualities the next generation must inherit.

Artificial Selection

In a herd of cattle, to banish or destroy the most promising individuals is to allow the inferior to become the parents of the next generation. This is race deterioration, although as heredity runs level, the individuals of the new herd must be on the average the equivalent of their parents. A scrawny herd is the natural offspring of scrawny parents. On the other hand, by preserving the most desirable types and these only, one gets a basis of continued betterment. It is said that when the short-horned Durhams first attracted general attention in England, the long-horned breeds which preceded them, inferior for beef and milk, vanished "as if smitten by a pestilence." Among animals as well as among men, the type is determined by the individuals which survive. This fact is the basis of the process known as Artificial Selection. By this process men have formed the various breeds of domestic animals and plants. The value and stability of these breeds depend on the preservation of the best for parentage.

Value of Individual Initiative

More vital than the elimination of the less advantageous hereditary elements in humanity is the encouragement and preservation of the more advantageous. It is this that counts most.

In all history, the influence of individual initiative has been a potent factor. In the long run it has been the most important element in the building of civilization. Great men have been "molders of environment" while the common man lies "at the foot of the strong god, Circumstance." With men in the mass, history repeats itself. The rare man of courage, wisdom and initiative interposes to prevent such repetition and a new epoch in human affairs is begun, a new type of history written.

In a scholarly work on the *Influence of Monarchs*, Dr. Frederick Adams Woods calls especial attention to the absurdity of speaking of the Romans, the Greeks, the French, as though these nations constituted a continuous series from century to century. In reality the history of these and other races is largely measured by the number and quality of the men of high ability produced among their number. And such men of ability are not descended from the mass but from superior strains, natural nobility within the rank and file of the race. Whatever cuts off its superior strains contributes to the downfall of the group containing them.

> "The Egyptians," says Dr. Woods, "probably never had any highly developed building instinct, though some of their rulers had. 'The Greeks' as a whole may never had been artistic and intellectual, though a percentage certainly were. 'The Romans' may never have had a special faculty for law and government. Such talents may have been confined to the patrician families....No matter what may be the form of government...the laws of heredity will work toward the formation of governing classes inherently superior to the sons of other men. Universal suffrage and universal education, the most carefully equalized scheme

of social opportunity cannot prevent this tendency of the homogeneous to pass into the hetrogeneous, – this splitting up of mankind into sub-varieties, castes and breeds.

"Historical science can scarcely at present predict the future, but it can interpret the past. If the work of the world has been initiated and directed by a few very great men, and if these men are the predetermined products not of outward but of inward differences, the true interpretation of history must hinge upon the gametes, and the laws which govern all organic life."

"In life or death," says Ellen Burns Sherman, "the man with presence of mind rarely counts as merely one. Indeed, Nature's basis of valuation of such men may be inferred from their numerical ratio to the rest of the population. For Nature, for her various ends has a human currency of divers metals and denominations. The ratio of Cæsars, Cromwells, Lincolns, and Shakespeares to the rest of the population has always been most economically predetermined."

Law of Quetelet

By the law of probabilities as developed by Quetelet, it is claimed that there will appear in each generation the same number of potential poets, artists, investigators, patriots, athletes and superior men of each type of excellence. But this law can hold only in case of absolute continuity of parentage. An equal percentage of persons carrying the hereditary potential for superior performance should survive to take the responsibilities of parenthood. Otherwise Quetelet's law, as Quetelet himself noted, becomes subject to the operation of reversed selection or the biological "law of diminishing returns."

Again, all laws of probabilities and of averages are subject to the primal law of biology, which no cross-current of life can overrule or modify, – "*Like the seed is the harvest*." There is a Moorish proverb which reads: "Father a weed, mother a weed, do you expect the daughter to be a saffron root?" One from the Spanish puts the obverse of this: "A lion breeds lions; a brave man has brave sons."

Captain W.A.C. Bowden-Smith

A regular army officer who was commissioned into the Royal Fusiliers from Sandhurst in 1909, Bowden-Smith took part in the Tibet Expedition of 1903-04, where he was wounded and received the King's Medal with Clasp. He was mortally wounded on August 23, 1914 at the Battle of Mons, dying at the age of 32, without children. A letter from a fellow officer reported that "Captain Bowden-Smith was very gallantly bringing up reinforcements at Mons, under fierce fire, when he was wounded in the right arm and abdomen. Out of the fifty men he brought up, all but three were shot down."

Workings of Primogeniture

The feudal nobility of each nation in Europe was primarily made up of the brave, the strong and the fair. By their courage and strength they became the rulers of the people, and by the same token they chose the fairest for their mates. In the organization of England particularly, the attempt was made to emphasize and perpetuate this superiority by the law of primogeniture. On "inequality before the law" British polity has always rested. A certain few have been fed on "royal jelly" as the young queen bee is fed, and thus raised to a higher class, distinct from the workers. To take this leisure class out of the struggle and competition of life, so goes the theory, is to make the first born and his kin harmonious and perfect men and women, fit to control the social and political life of the state.

Primogeniture has brought, however, a real gain to the nation though not to the individual. This lies in the fact that the younger sons and the daughters' sons were forced constantly back into the mass of the people, insuring among them the presence of dominant strains. Englishmen of today are descendants of the old nobility, and in the stress of natural selection they have crowded the children of the swineherd and the slave. Younger sons manned the *Mayflower*, and in each generation they have striven for liberty in England and the United States. Studies in genealogy show this to be literally true. All the "old families" in New England and Virginia trace their lines back to nobility and hence to royalty. Indeed, almost every Anglo-American has, if he knew it, noble and royal blood in his veins. The Massachusetts farmer, whose fathers came from Devon or Somerset, has as much of the blood of the Plantagenets, of William and of Alfred, as flows in any royal veins in Europe. But his ancestral line passes through the working and fighting younger son.

On the continent of Europe the law of primogeniture was little emphasized. There the nobility formed a distinct class, all of noble blood being included. Gentle blood rarely mingled with that of the commoner. Noblemen were brought up in luxury, their maintenance laying on ever increasing burden on the villager and the farmer. In France, the intolerable load of taxation contributed to the Revolution with its Reign of Terror, and the sacrifice of the "best that the nation could bring," for the victims numbered many of the best from the standpoint of race value. Their weaknesses were those of softness induced by indolence a defect not genetically transmissable. Under other conditions, they might have defended themselves more effectively.

Racial Loss Through Emigration

Emigration has played a large part in the depletion of peoples in different districts of Europe and even in older sections of the United States. This may mark a loss to a particular region, but none to the world, the value of a man and his posterity, broadly speaking, being as great in one place as in another. Moreover the pioneer gains by travel, picking up something on the road, though he may also lose through separation, as in the new freedom he tends to fall out of touch with the achievements of the older social fabric. But on the other hand, he will escape many hampering traditions, and the sturdiness of racial stock is only slinghtly influenced by culture, the social values of native strength reasserting themselves when opportunity offers. Meanwhile the gains in the New World may be traced as losses to the Old. For example, from the counties of Devon and Somerset arose, primarily, the colony of Massachusetts Bay. From the loins of Old England, New England arose, and from self-governing New England, the democracy of the United States.

In the days of Queen Elizabeth no county in England had more abounding life than Devonshire. "The Dragon Persecution" there, and the "Bloody Assizes" in Somerset, sent them forth as emigrants. Devon now lives across the sea, whither three hundred years ago her young men carried the venerated names of her picturesque seaports.

Racial Loss Through Immigration

By *Immigration*, lands scantily occupied by barbarous races have been replaced by peoples more efficient or more aggressive. Through the same agency strong nations have sucked in weaker groups to fill the vacuum caused by war or to meet the demands of industry. Through conquest by war as well as out of industrial needs grew up the institution of slavery. In Rome, "whole tribes were borrowed" for the work of agriculture,

Major J.S. Cawley

A regular army officer who was commissioned into the 20th Hussars, and served in South Africa, where he received the Queen's Medal with four Clasps and subsequently in Egypt, and India. Major Cawley went to France with the 1st Cavalry Brigade, and was killed by a shell while trying to restore order following a stampede of artillery horses under heavy shell fire on September 1, 1914. He died without issue, aged 34. His elder brother was killed at the Battle of the Dardenelles on September 23, 1915.

while conquered groups were utilized as menials or slaves.

Everywhere, under these conditions, the blood of the slave or the conquered has diluted that of the dominating race, usually to its detriment. For example, in most Spanish and Portugese Colonies, Latin blood has been mixed with aboriginal, producing crosses showing few of the virtues of the European stock. Indeed, in Portugal, the mixture from subject races in Brazil, Africa and India has invaded the parent itself to its social and political confusion.

Two main facts appear in this connection. In many racial crossings occurs the mingling of the least desirable types of each. Naturally where the dregs of one race mix with the off-scourings of another arise distressing possibilities of vice and incompetence. For instance, the Eurasian in Asiatic seaports "is damned from his birth and on both sides."

The words "hybrid" or "mongrel," terms of reproach as usually applied to the human race, related commonly to the union of widely different peoples. But the question of "Race or Mongrel" cannot be settle by *a priori* assertions as to superiority of pure over mixed races. There is no general law that mongrels are sterile, and non-resistant – or the reverse. No useful information can be gained from the study of the first generation. One would need to know the nature of the recessive characters involved as well as of the dominant ones. The final Mendelian disposition of mixed race characters must determine the ultimate answer.

The intermarriage of European peoples can hardly be called crossing at all, as the racial differences concerned are of slight order, little more than temperamental at the best, and most of the traits we commonly recognized are matters of education. All those qualities which disappear in a generation in America must be chargeable to education, not to race. There is no widespread race which is pure. There is no such thing as a pure-blooded German or Frenchman. "Saxon and Norman and Dane are we" of England. Likewise are we Briton and Welsh and Cornish; also Scotchmen, Highland and Lowland, Manxmen, Ulstermen and Irishmen.

That the crossing of the closely allied European peoples in America has, of itself, brought no disaster to our republic is a matter of visible observation. That wide crosses necessarily work for evil in not proved. But we may admit that the introduction of African blood has not been a gain to the republic. And we may also admit that much of later immigration from Europe and Asia has lowered our average. America was a haven of refuge from senseless tyranny. Immigration thus brought to the new world a wealth of initiative and adaptability such as no nation ever inherited before. But in later days this current has changed. Wider opportunity has opened before the common man in the more progressive nations, and the incentive of freedom has been less acute. Among immigrants to America today are some with magnificent personal responsibilities, men of the stuff that makes republics. But the most of them are not such, and while their presence adds to our material wealth they constitute, as a whole, a burden on our democracy. Only a man who can take care of himself and have something left over for the common welfare is a good citizen. It is hard to maintain the principle of equality before the law among people who have never felt and never demanded such equality.

Racial Inequality

The claim is sometimes made on an assumed basis of science that all races of men are biologically equal, and that the differences of capacity which appear are due to opportunity and to education. But opportunity has come to no race as a gift. By effort it has created its own environment. Powerful strains make their own opportunity. The progress of each race has depended on its own inherent qualities. There has been no other leverage. Physical surroundings have played only a minor part. To say that one race as a whole is inferior to another is only to repeat what is said every day of individual men. This does not imply that a lower race may not produce its own prophets or scholars or heroes. That race is lowest which shows, on the whole, least capacity for self-elevation.

"All men are born free and equal," it is asserted, but such equality is political only. It cannot be biological. In every race are certain strains having capacities not attainable by the mass. There may be equality of start, equality before the law, but there will always be differences of attainment. The gifts of potentiality, unit characters of the germ-plasm, are not equally shared by all people of the same race. The average status of one may be below that of another, and the highest possibilities of

Captain A.R.M. Roe

A regular army officer who was commissioned into the Dorsetshire Regiment in 1900, Captain Roe joined the First Expeditionary Force in France in August 1914, and on September 9, during the advance on the Aisne, he was wounded in the arm while leading his Company against a German position. Despite his injury, he continued fighting until he was also hit in the head. Seven days later he died of his wounds, aged 32. Captain Roe was married in 1911, and left a widow and one daughter.

one type may be greater than those of another. In general, the highest range of possibilities in every field has been reached by the "blonde races" of Europe. Groups of less individual or of less aggregate achievement may properly be regarded as "lower."

(1) Εὐγενεία, the condition of being well born; ὐθενεία, the condition of being well brought up.

(2) "Nature is obstinate and will come running back even though you expel her with a fork." (Carlyle.)

IV. WAR AND RACE SELECTION

Franklin's Views

Benjamin Franklin who, above all his contemporaries, "saw through the forms of things and laid bare the substance" was, so far as records go, the first man in all history to notice the necessary relation of war to the breed. Of an interview he had with Franklin in 1783, John Baynes records the following:[1]

"Insensibly we began to converse on standing armies, and he, seeming to express an opinion that this system might some time or another be abolished, I took the liberty to ask him in what manner he thought it could be abolished; said he: 'I admit that if one Power singly were to reduce the standing-army, it would be instantly overrun by other nations; but yet I think there is one effect of a standing-army which must in time be felt in such a manner as to bring about the total abolition of the system.' On my asking what the effect was to which he alluded, he said he thought they diminished not only the population, but even the breed and the size of the human species. 'For,' said he, 'the army in this and every other country is in fact the flower of the nation – all the most vigorous, stout and well-made men in a kingdom are to be found in the army. These men in general never marry.'"

"History," observes Dr. James Brown Scott, "is but a commentary on the statement of Dr. Franklin, for standing armies and their destruction in battle have sacrificed the fit to the unfit and ruined the nation on the battlefield. We may close our eyes to history and refuse to listen to its teachings, but the fact is and always has been that war deprives a nation of the most fitted to maintain its existence, and a succession of wars ruins the stamina of a nation, no matter by what sophistry we may disguise the fact or explain the consequence.

"It is not maintained or asserted that war may not draw out the higher instincts of a nation, that courage and self-sacrifice, of which we are proud and whose traditions we cherish, are not produced and made prominent in war in ways impossible in peace, but the misfortune and the scourge of war lie in the fact that these very qualities are sacrificed and lost."

The late Professor Jakov Novicow of the University of Odessa, one of the most learned and vigorous opponents of war, has said:[2]

Lieutenant G.L. Relton

A regular officer who was commissioned into the East Surrey Regiment shortly before World War I and who proceeded directly to France with the outbreak of hostilities. He received three bullet wounds on September 14, 1914 in the Battle of the Aisne, and died shortly thereafter, aged 23, without issue.

"War produces indeed a selection, a choice of the worst. The young men strongest and most healthy go to the war. Among its combatants, the most valiant take the lead. In consequence, the more perfect the individual, the greater his chance to be killed. In most battles it is the best that fall. On the other hand, the feeble and sickly elements, those not enrolled under the banners of war, reproduce themselves, while the flower of the nation is condemned to celibacy or to relations with prostitutes, this leading so often, alas, to the most fatal results."

Plainly, the history of peace in the various nations is the history of the slow recuperation of races from the effects of war. Most phases of natural selection, in society as in nature, make for advance, very slow no doubt, but real. Every war means a step backward, long or short in proportion to the accompanying destruction of virile elements.

Professor Charles Richet,[3] of the Chair of Physiology in the University of Paris, thus discusses military selection:

"In nature, when two animals fight with one another, it is the more valiant that survives. Disease attacks the weak ones; those of greater vigor and courage live on to perpetuate the race of the courageous and strong. But in war among men...the selection is reversed, and conduces to the impoverishing of the race.

"First, let it be remembered that the sick and the infirm are exempt from service. Those who have any weakness are well protected by the military laws, and not one of these unfortunates runs any risk of perishing on the field of battle. Those who are chosen to disappear are the halest and heartiest. Robust youths, the hope of future generations, these are the ones who are declared fit for service.

"On the field of battle it is always the same kind of men," according to a well known saying,[4] "who get themselves killed."...

"From a biological point of view, long wars are exhausting to a nation and conducive to actual degeneracy. For at last, as during the period of the Napoleonic massacres between 1798 and 1815, all the able-bodied population ends by being annihilated on the field of battle, and the weak and infirm and cowardly are alone left to carry on the race. This is one of the most serious evils among the innumerable ones which follow in the wake of war."

War as Race Suicide

"The Great War," says Romain Rolland, is "a sacrilegious conflict, which shows a maddened Europe ascending its funeral pyre, and like Hercules destroying itself with its own hands."

"Each nation justifies its own share in the present struggle on the ground that it is virtually waging a war of self-preservation. If all this is the outcome of a war of self-preservation, one would like to know what form a war of self-destruction would take. 'Your King and country need you' is the patriotic appeal, and those who respond are immediately thrust by King and country as fuel into a smoking furnace...."[5]

Says Professor Kellogg, "War to the biologist seems, above all else, stupid. It is racially dangerous. It so flies in the face of all that makes for human evolutionary advance, and is so utterly without shadow of serious scientific reason for its maintenance. It is not natural selection in Man, nor in any way the counterpart of it. Its like does not exist in Nature outside the forays of the few degenerate fighting ant species, some of whom have lost the power of caring for their young, and hence live as social parasites on less barbarous kinds, or have given up all other means of food getting than robbery by force of numbers. It is not only not natural selection, but its results are an artificial, unnatural reversed selection, one that turns on itself, giving no advantage to the conqueror, but only many and terrible disadvantages to victor as well as to loser."[6]

In a recent address in Worcester, Massachusetts, Dr. G. Stanley Hall, President of Clark University, thus speaks of the anti-eugenic aspects of war:

"Seven to ten millions of the soldiers now in the war, or training for it, are married men and are the most able-bodied and intelligent potential fathers. Statistics can tell us approximately how many children would, on the average, have been born of these men, had they stayed at home....Thus the crop of best babies, which is the most precious of all assets for both national and cultural prosperity, and on which national greatness depends more than upon anything else, is greatly reduced, for, to say nothing of the killed who will never be parents, we must also consider the vastly greater number who, as medical studies of the effects of war show, suffer impairment in the quality of their future parenthood..."

Major W.C. Christie

Commissioned into the Royal Warwickshire Regiment in 1891, with whom he participated in the Battles of Atbara and Omdurman, Major Christie was mentioned in despatches and received the Queen's Medal, the Khedive's Medal with two Clasps, and the Order of Medjidieh. He served in the South African War at Vet River, Zand River, Diamond Hill and Belfast, being again mentioned in despatches and being awarded the Queen's Medal with five Clasps and the King's Medal with two Clasps. Joining the Western Front with the 1st Army Corps in August 1914, he was twice mentioned in despatches at the battles of the Marne and the Aisne, and died leading his men in an attack at the village of Meteren on October 13, 1914. A fellow officer wrote: "he was a most gallant Officer, a man with absolutely no sense of fear . . . he could never do anything small or petty." Another wrote: "A more gallant Officer and gentleman never breathed . . . we got our position (in the attack) chiefly owing to Christie's gallantry . . . I heard one of the men remark, 'He's a daring man. I hope we don't lose him.'" A soldier groused "it's this - - - - Christie. We are always in the thick of it with him, and worst of it is, there's not one of us as wouldn't go to 'ell with him willingly." Major Christie left only one child.

Captain R.C. Eveleigh

A regular officer who was commissioned into the Oxfordshire and Buckinghamshire Light Infantry, Captain Eveleigh served in Mauritius before joining the First Expeditionary Force in the battles of Mons, the Marne and the Aisne. On September 16, 1914, despite wounds he led a small group to save sixteen wounded German soldiers from a burning farmhouse. Two days later he was killed by a shell splinter, while seeing his men into shelter under heavy artillery fire. Aged 29, he left no issue.

Elbert Hubbard observes: "The man who escaped conscription through weakness (marches his) disabilities down the winding ways of time."

"Europe must breed," says Bernard Shaw, "from the men of the last reserve."

"The strongest and best men," says Robert L. Duffus, "those fittest to be the fathers of coming generations, are picked out ...to be mowed down by shell, to be weakened by hardship and overstrain, to contract and perhaps to pass on to other generations the hideous diseases of camps. Every great war leaves the general average of health, strength, intelligence and morality a little lower."

Dr. Caleb Williams Saleeby of London, discussing *The Long Cost of War*, says:[7]

"We all find reasons for the fall of the Roman Empire according to our creeds, instincts and prejudices. But some of the reasons advanced actually have reason in them. The incessant drain of the right kind of military stuff from the population of Rome, led in the long run, to the production of that degenerate people who wish only for *panem et circenses* (bread and circuses). The recruiting officer rejected the halt and blind, feeble-kneed, the easily fatigued, saying, though he did not know it: 'You are not good enough to be a Roman soldier; stay at home and be a Roman father.' The future was ruthlessly sacrificed by militarism to the present, even as now in Northern Europe.

"This very morning as I write, comes the news that several famous athletes of our own race have been killed in France. They may have been winning or losing, retreating or advancing, but they are dead, and Britain will have no more sons of theirs. Similarly the correspondents tell us how, in Paris and elsewhere, none of the able-bodied remain, between, it may be, the ages of nineteen and fifty. How, then, is the race being recruited, while the regiments are being recruited? With some personal knowledge of and almost boundless admiration for France, I can scarely doubt that the heaviest burden under which France now bows is the lack of those sons of hers whose grandfathers that should have been, fell a century ago...."

War and Selection

To demonstrate the power of war selection let us consider a quality which is not of great importance in itself, and that is stature. At the time of the Napoleonic Wars, European armies sought tall men because they looked more impressive on the battlefield.

The lowering of the stature of the French soldiers due to the wars of Napoleon has become a matter of common knowledge. It is the saying in France that "Frenchmen are small because all our tall ancestors died in our victorious wars." Legoyt thus states the case: "It will take long periods of peace and plenty before France can recover the tall statures mowed down in the wars of the Republic and the First Empire."

It should be clearly noted that a mere decline in stature is in itself of little racial significance, save as an index of decline in other and more vital regards. Tall stature has been sought for in recruiting armies and so have qualities of boldness and dash. The decline in stature can be measured; other qualities cannot, but we may fairly assume that all soldierly traits have suffered together and the measure of one serves in some degree as the measure of all.[8]

This matter has been made the basis of a critical study by Professor Vernon L. Kellogg, in the interests of the Carnegie Endowment for Peace. A synopsis of the results of this study is given in *Social Hygiene*, December, 1914.

"Running nearly parallel with the fluctuation in number of exemptions for undersize is the fluctuation in number of exemptions for infirmities. These exemptions increased by one-third in twenty years. Exemptions for undersize and infirmities together nearly doubled in number. But the lessening again of the figure of exemptions for infirmities was not so easily accomplished as was that of the figure for undersize. The influence of the Napoleonic Wars was felt by the nation, and revealed by its recruiting statistics . . . ."

Ammon's Arguments

The most important attempt in the name of science to minimize the evil effects of military selection is that of Dr. Otto Ammon of Jena.[9] His special thesis is that the best men

Major E.A.C. Blake

Major Blake was a professional army officer who was commissioned into the Durham Light Infantry in 1896, and served in the South African War, being present at the relief of Ladysmith and the battles of Colenso and Spion Cop. He was severely wounded, mentioned in despatches, and awarded the Queen's Medal with two Clasps and the King's medal, also with two Clasps. Arriving in France with the First Expeditionary Force, Blake was killed by artillery fire on October 20, 1914 near Lille, after surviving the battles of Mons, the Marne and the Aisne. He left no children.

Captain H.H. Kelly

A regular army officer, who was commissioned in 1899 into the Royal Engineers, Captain Kelly served as an Engineer with the Egyptian Army, and became Inspector of Roads and Communications for the Sudan, and Commissioner for the Sudan-Uganda Boundary Commission in 1913. He took part in various military operations in the Sudan, and carried out explorations in Abyssinia, being elected a Fellow of the Royal Geographic Society. With the outbreak of war he went to the Western Front and was killed by a sniper while superintending the erection of barbed wire defenses on the night of October 24, 1914. He died, aged 34, without issue.

should rule, and that such a condition is brought about by selection ruthless and without regard to any principle of equality or democracy. He contends that war has this result, exerting on the whole a helpful and advantageous selection. He admits that partisan warfare, banishments and executions in ancient time left Rome weak in men of force and ability to lead. But a different result comes from the massacres in war and conquest. In them, it is the weak rather than the strong who are extirpated. "A hundred years after the unspeakable desolation of the Thirty Years' War, in which Germany was robbed of three-fourths of her people, arose Goethe and Kant." Thus while recognizing, in part at least, the fact of reversal of selection by war he denies it in the case of massacre.

This paper of Dr. Ammon, with many similar discussions by authorities in Germany, is vitiated through the omission of a single fact. A selection much to the detriment of the population had been made already before war, the strong being chosen for military service and for decimation while the weak were released. The army had no use for the cripples, the deformed, the organically diseased and the dwarfs. These were all handed back to the population by the recruiting commission.

Losses in War

Colonel G.F.R. Henderson of London thus estimates the losses in killed and wounded in twenty battles showing greatest fatalities, between 1704 and 1882.

Battle	Date	Number Engaged	Killed and Wounded	Percentage
Leipzig	1813	440,000	92,000	20
Borodino	1812	262,000	75,000	28
Aspern	1809	170,000	45,000	26
Wagram	1809	370,000	44,000	11
Eylau	1807	133,500	42,000	33
Waterloo	1815	170,000	42,000	24
Gettysburg	1863	163,000	37,000	24
Chickamauga	1863	128,000	35,000	27
Friedland	1807	142,000	34,000	23
Malplaquet	1709	200,000	34,000	17
Vionville	1870	168,000	32,800	19
Zorndorf	1758	84,760	32,000	38
Solferino	1859	295,000	31,500	10
Blenheim	1704	116,000	31,000	26

Kunnersdorf	1759	113,000	31,000	27
Gravelotte	1870	320,000	30,000	9
Königgratz	1866	317,000	26,894	6
Wilderness	1864	179,000	26,000	14
Austerlitz	1805	148,000	25,000	16
Spottsylvania	1864	150,000	25,000	16
Ypres[10]	1915	1,000,000	500,000	

According to the *London Evening News*, the latest list (173) in the Prussian Armies make a total of killed, wounded and missing (to March 1, 1915) of 1,050,029 men. Besides these, 160 lists from Bavaria, 136 from Würtemberg, 119 from Saxony and 20 from the Navy, 435 in all, have been issued in Germany.

The *California Outlook* thus enlarges on these estimates: "The last eight Prussian lists contain 33,142 names, or an average of 4,143 to each list. The average for the entire 173 Prussian lists is 6,069 to each list. Taking the smaller of these two averages, and cutting even it down to 4,000, by striking out the odd numbers, and we would have, even on this underestimate, 1,740,000 casualties for the non-Prussian part of Germany, or 2,790,029 for the entire German army. This is more than all the soldiers who served for any period, short or long, on the Union side in the American Civil War. Taking the average of the Prussian lists as probably the average for the others, we have 2,640,015 German casualties outside of Prussia, or 3,690,044 for all the German armies. This is more than all the soldiers on both sides in the American Civil War.

"In other words, if every soldier who enlisted in our whole four years, had been lost, the number would not yet have equaled the losses of the German army alone, in a little over half a year of war. It would certainly be an underestimate to compute the Russian, Austrian, Belgian, French and British losses combined at twice those of the Germans. Probably they are nearly proportionate to their total numbers engaged. But even at three times, the losses to date of all the European armies must be over ten million men. There are about twenty million adult, able-bodied men in the United States. The European losses already equal half of these — a sum of grief and tears and blood equal to depriving every second family in America of its adult father or adult son. It is a thing colossal, staggering, incomprehensible. Nothing remotely approaching it ever happened in the world before, and civilized mankind could not survive it happening twice."

Major W.L. Allen, D.S.O.

A regular British Army officer who was killed on October 25, 1914 at Ypres. Major Allen was commissioned into the King's Own Yorkshire Light Infantry. He served in the South African War, being present at the battles of Colenso, Spion Kop, and Pieter's Hill, winning the D.S.O., the Queen's Medal with five Clasps, the King's Medal with two Clasps, and being twice mentioned in despatches. Having reached the age of 43 before the war began, he was among the minority of those that died in World War I who left children behind them.

Lieutenant J.H. Fraser

The only son of a British colonial official, Lieutenant Fraser was commissioned as a regular army officer from Sandhurst in 1906 and served with the Gordon Highlanders in India, winning prizes for military engineering. Going to the Western Front in October 1914, he was killed on October 30, 1914 near Ypres. Mentioned in despatches, he was described as "one of the finest Officers I have ever met, absolutely fearless, always cool, and every soul in the Battalion admired him for his splendid soldierly qualities and loved him for his character." He died, aged 26, without issue.

Are There Compensations?

In every war, it has been argued that there have been certain compensations in exaltation of character among soldiers in the field and, quite as often, among those who suffer at home.

In an eloquent paragraph, in which he lays great emphasis on the biological evils of reversed selection due to modern war, Professor J. Arthur Thomson of Aberdeen adds this note of appreciation of the soldier: "Who does not admire what Mr. Sandeman says in his *Uncle Gregory?* That quite unmistakable note that you get in a very few people who in one way or another have actually accepted death and are only, so to speak, alive in the meantime. It belongs to the flawless perfection of the military spirit with its entire detachment from life itself, from self-will, from fear, from ease and from all pretense."

The son of a valued friend in Germany wrote this to his sister just before his death in Poland: "One who stands in the field, so often face to face with death, knows how to value life. But he loses also the fear of death, for he knows that the highest fortune is the forgetting of personality, the offering up of self. And this takes all terror from death."

"In Europe, in war time, moral conditions," says Julia Grace Wales,[11] "are very far from normal – not abnormally low, but almost superhumanly high. The very unity and cohesion of a race has carried the individual beyond his normal range. Each people is as a single family; there is neither high nor low, rich nor poor, but a brotherhood of men. No man counts his life dear unto himself. All are fighting, with unquestioning devotion for homes and fatherland, for language, institutions, traditions, for all that they hold most sacred and most dear. Whatever we may believe about the folly or the deliberate wrongdoing of governments, the fact remains that each people is in a state of spiritual exaltation. Individuals are everywhere thinking, feeling, suffering, facing the ultimate issues of life and death. Their senses are sharpened, their spirits sensitized to the significance of what had become commonplace, to familiar landscapes, to the assocations of home, to the ideals of the race, to its heroism and its poetry, to the symbols of its religion. This thing is like a tidal wave of the sea; it has drawn deep."

In his charming studies of *Feudal and Modern Japan*, Mr. Arthur Knapp, then of Yokohama, returns again and again to the great marvel of Japan's military prowess after more than two hundred years of peace. This was demonstrated in the Chinese war, and more conclusively shown on the fields of Manchuria since Mr. Knapp's book was written. It is astonishing to him that, after more than six generations in which physical courage had not been demanded, this virile virtue should be found unimpaired.

But this is just what should have been expected. In times of peace there is no slaughter of the strong, no sacrifice of the courageous. In the peaceful struggle for existence a premium is placed on vigor and intelligence. "What won the battles on the Yalu, in Korea or Manchuria," says Professor Inazo Nitobe, "was the ghosts of our fathers guiding our hands and beating in our hearts. They are not dead, these ghosts, these spirits of our warlike ancestors. Scratch a Japanese, even one with the most advanced ideas, and you will find a Samurai." Translated from the language of Shintoism to that of science we find it a testimony to the strength of race-heredity.

If after two hundred years of incessant battle Japan still remained virile and warlike, that would indeed be a marvel. But such marvel no nation has even seen. It is doubtless true that warlike traditions are most persistent with nations most frequently engaged in war. Traditions of war, however, and the physical strength to gain victories are very different.

The little mountain kingdom of Montenegro is sometimes used as a demonstration that much war breeds strong men, for the Montenegrins, physically superior to their brother Serbians, are tall, straight and strong and they have fought much, within and without, as their belts crammed with daggers and pistols seem to testify. Time and again have they made forays on the Turks in Albania, taking their fortresses by storm, while at home the custom of private revenge for personal wrong has been a duty almost religious. Herbert Spencer quotes from Boué: "In Montenegro, one will say of a man whose clan has killed a member of another, 'This clan owes us a head and this debt must be paid, for who does not revenge himself cannot be sanctified.'"

Originally the Montenegrins were Serbians who defied the Turk and fled to their inaccessible linestone crags. They were a picked groups physically, men of indomitable will. They have not mixed with other races, and all their members have practi-

Lieutenant A. Gilliat-Smith

The only surviving son of his parents, Lieutenant Gilliat-Smith was commissioned into the Royal Engineers to join the First Expeditionary Force in France. He was killed on November 1, 1914 near Ypres, aged 26, while endeavoring to get his men under cover from heavy cross-fire. He left no children.

cally the same inheritance of superior blood. They constitute a little group selected for courage and not yet ruined by war.

Ruskin's Testimony

John Ruskin once gave an address on the higher ideals of war before the cadets of the Royal Military Academy at Woolwich.[12] He spoke of the nobler features of the profession, its freedom from selfish ends, its "cleanliness" as compared with the soot of industrialism.

"All pure and noble arts of peace," he said, "are founded on war; no great art ever rose on earth but among a nation of soldiers. There is no art among a shepherd people if it remains at peace. There is no art among an agricultural people if it remains at peace. Commerce is barely consistent with fine art, but cannot produce it. Manufacture not only is unable to produce it, but invariably destroys whatever seeds of it exist. There is no great art possible to a nation but that which is based on battle."

The above words have been many times quoted, not for their inherent value, but because Ruskin wrote them. They constitute a sort of paradox, for everywhere else, even in the same address, their author accentuates the value of human life. For the time being, his mind is not fixed on the actualities of war. Every form of human activity (unconnected with coal and steam), war included, might contribute, he believed, to spiritual elevation. But he was plainly not thinking of the warfare in which these cadets might have to engage; rather of the ancient contests between man and man. Of modern slaughter by machinery elsewhere he says:

"If you have to take away masses of men from all industrial employment – to feed them by the labor of others – to move them and provide them with destructive machines, growing daily in national rivalship of inventive cost; if you have to ravage the country which you attack – to destroy for a score of future years its roads, its woods, its cities, and its harbors; – and if, finally, having brought masses of men, counted by hundreds and thousands, face to face, you tear those masses to pieces with jagged shot, and leave the fragments of living creatures, countlessly beyond all help of surgery, to starve and parch, through days of torture, down into clots of clay – what book of accounts shall record the cost of your work? That, I say, is modern war, – scientific war, – chemical and mechanic war, worse even than the savage's poisoned arrow."

Do We Exaggerate?

Turning for a moment to the war now raging in Europe, certain final effects may be discerned. As the greatest of all wars, it must prove the most disastrous, and such from every point of view. The number of losses already (March, 1915) rises high into the millions. All these individuals had been selected for vigor and strength. The various armies engaged include the great body of the university men, athletes and skilled laborers[13] in each of the belligerent nations. The conditions of this war leave little hope that any large percentage of those on the firing line will return unscathed. The future will show, doubtless as never before, and in all nations alike, that war-selection points the way downward. To what degree this will be felt and what will be its visible effects on society, we have no precedent by which we may adequately estimate. That the damage will be greater in fact than will show on the surface one may be very sure. It is the men of initiative who mould civilization. Through them social and political betterments arise. "An institution is but the lengthened shadow of a man." The nation which lacks epoch-making men will fail in epoch-making history.

It will no doubt be said by some who read this book that its thesis is an exaggeration, that war is but one influence among others which sift the human breed for better or for worse and that all forms of destructive selection nature provides an antidote. There is in certain quarters, especially in England, a curious perversion of ideas of cause and effect. It is freely admitted that war does indeed destroy many of the best, leaving to a large extent the second-best to sire the coming generations. But we are further told that this defect is to be remedied by compulsory military drill. The weak will then be made virile and capable of begetting vigorous progeny. There is not much truth in the first assertion and none in the second. No results of a process of this kind on the part of the male parent have permanence in heredity. Men inefficient by nature have progeny of like type, and the case for the children is not materially mod-

Lieutenant N.W.A. Henderson

A regular army officer, Lieutenant Henderson was commissioned in 1911 into the Royal Scot Fusiliers, and went to France in August 1914. He was killed in action on November 10, 1914, age 23, leaving no issue.

Second Lieutenant L E.F. Grubb

An Exhibitioner of Brasenose College, Oxford, Lieutenant Grubb enlisted at the outbreak of World War I and arrived in France in time to participate in the retreat from Mons, during which he served as a despatch rider. Commissioned early in November 1914, he was killed on November 14, 1914, leading his platoon in an attack on an enemy sniper position. He died aged 22, without issue.

ified by a superficial alleviation of the physical limitations of the father.

The young men in the British cadet corps seem stronger than those outside, because they are selected from the beginning. No officer wants a recruit who cannot be made to look well in uniform. General Ian Hamilton is reported to have said that if the War Office had control, "never for one moment would a soldier Inspector-General endure the tens of thousands of weak eyes, incipient deafness, hammer toes, flat feet, knock-knees, now disfiguring our elementary schools." (Impey.)

But these defects are hereditary qualities, the legacy from previous generations of just such people, rejected from the armies by drill-sergeants of the past. There may be alleviation for them, by surgical or other methods, but they cannot be eliminated by military drill applied to those who have never suffered from such defects.

"The facts speak for themselves. Serious war and the preparedness for serious war mean the temporary or permanent withdrawal from the population of a part of it selected for physical vigor and often for courage, patriotism, and idealism, and the exposure of this part to special danger from death and disease. This death and disease, under the circumstances, are not race-purifying or race-enhancing, but race-deteriorating, through the encouragement of poor breeding. Every race needs its best possible inheritance. Any institution that tends to give it less than that is a race-injuring institution."[5]

"The most economical and most positive factor in human progress," says Professor Kellogg, "is good breeding. Race deterioration comes chiefly from its opposite, bad breeding. Militarism encourages bad breeding. Despite all delusive phrases to the contrary, the maintenance of any army is a preparation for war and a step toward war and not toward peace. Do governments, or will they, maintain this blessing of military service for the health and eugenic advantage of their people? Is it not done solely from the stimulus of expected war? Is it not done solely with the full expectancy and deliberate intention of offering this particularly selected and cared for part of the population to the exposure of wholesale mutilation and death? And this death is to come, if at all, before this extra-vigorous part of the population has taken its part in race propagation, the precise function the performance of which the race most needs from it."

(1) *Memoirs of the Life of Samuel Romilly*, by his sons. 2 vols. 1842. vol. I, p. 69, quoted by James Parton and by James Brown Scott.

(2) The following paragraphs are condensed from Novicow's *La Critique du Darwinism Social.*

(3) *Le Passé de la Guerre et l'Avenir de la Paix.*

(4) "À la Guerre, ce sont toujours les mêmes qui se font tuer."

(5) *Topeka Journal.*

(6) *Beyond War.*

(7) *Westminster Gazette.*

(8) "Just off the southern shore of Newfoundland lie the islands of St. Pierre and Miquelon – rather pathetic vestiges of the broad domain France hoped to possess on this western continent. The simple folk in this retired colony have been asked to send 800 men to the war. When the physical examinations were over, the officers found that just 300 men were fit for service. Those chosen men, the very pick of the colony, its best hope for growth and betterment for generations to come, are now at sea on their way to the front. Before them is desperate service for their mother country, behind they leave the chaff and riff-raff from which they were winnowed. This instance of destructive selection takes our attention because it comes close to our own shores. We should remember that through the villages and towns and cities of almost all Europe the same process is at work. The war is gnawing at the vitals of our race like the vulture that tortured Prometheus." (Edwin D. Mead, in the *Boston Herald.*)

(9) *Die Gesellschaftsordnung und ihre Natürlichen Grunlagen.* Jena, 1896.

(10) Estimate of Will Irwin.

(11) "The Wisconsin Plan" of Peace.

(12) *The Crown of Wild Olive.*

(13) It is said that in Germany 3,500 of the best picked mechanics have been assigned to the Aviator corps.

Captain W.M. Burt-Marshall

A professional soldier who was commissioned into the Argyll and Sutherland Highlanders, Captain Burt-Marshall joined the 19th Infantry Brigade in France in August 1914, and fought in the Battle of the Aisne. After lying all night in a wood under heavy shell fire, Captain Burt-Marshall led a bayonet charge: "he was at the head of his men and led the charge right up to the German barbed wire, was hit and fell, but rose again and dashed on to the trench, where he fell again. No one could get up to him, and those who were able crawled back to reform with the remnants of the Company." Captain Burt-Marshall died of his wounds in a German Field Hospital on November 17, 1914, aged 27. He left no children.

V. WAR AND WOMEN

Selection Among Women

We may here note that the process of selection in the course of war is confined mainly to men. If the fittest among the women were also destroyed, the proportion of decline would be twice as rapid. Women – personally the greatest sufferers from war – in a measure save the day because they are not subject to the reversal of selection. "Yet one consoling fact," observes Dr. Saleeby, "alone prevents this longest price of war from ruining even victorious nations more quickly and surely than it does. It is that war does not demand the healthiest and bravest of a nation's womanhood to be destroyed for the glory of the men who make wars. At least the generation to come may have mothers and grandmothers as fine as if there had been no war at all, and, of course, so impartial are the laws of heredity, both boys and girls to come profit accordingly." Unfortunately he is wrong. The best of the women who survive often fail to find fitting mates and so remain childless.

Womanhood and War

"It is especially in the domain of war," says Olive Schreiner (*Woman and Labor*), "that we, the bearers of men's bodies, who supply its most valuable munition, who, not amid the clamor and ardor of battle, but singly and alone with a three-in-the-morning courage shed our blood and face death that the battlefield might have its food, more precious to us than our hearts' blood; it is we, especially, who, in the domain of war have our word to say, a word no man can say for us...Only a woman knows what a man costs."

Dr. Anna Garlin Spencer[1] discusses effectively the reasons why women should hate war as the chief enemy of womankind. She says:

"Women bear the chief burden of personal care of the young, the undeveloped, the frail and sick, the aged, the feeble-minded, the socially incompetent. They have had to bear that burden ever since social sympathy forbade the strong to kill the weak by fiat of the state. This process of social protection of the incompetent has unquestionably lowered the average standard in human quality where it has worked unmodified by some science and ar of race culture. War – and all that makes for war – is the worst hindrance to the attempt to relieve women of this overmastering burden of administering philanthropy, and to give her time and opportunity for her organic function of teaching and developing the normal and super-excellent specimens of the race. Not only does it destroy uselessly all the common wealth of humanity so terribly needed for projecting and realizing the social control that can truly advance individual life, but it deliberately and monstrously aids that 'breeding downward' which is the bane of civilization....

"It is because of women's peculiar functional relation to the social demand for race integrity and race culture that enlightened women must hate war and all that makes for war. It sinks under waves of bestiality and passion those ideals on which respect for womanhood and tender regard for the child have fibered the later progress of the race."

"War Brides"[2]

At the beginning of the present war efforts were made in various nations (Germany, Great Britain, Austria, Turkey) "to guard against a falling birthrate" by offering special inducements to marriage before leaving for the front. This course has as a result the dubious advantage of making maids into widows and leaving them to bear children under great nervous stress. I am told that in Berlin, in early Ausust, 1914, more than 50,000 of such marriages were celebrated. A similar kind of war-mating took place in many other military centers.

In favor of this arrangement it has been maintained in England that for many men already engaged, it made marriage possible and compatible with enlistment. In a certain number of cases, this was no doubt true. It is further urged that for two or three million of women in Europe, matrimony must otherwise be wholly impossible. 'Better a day of wedded life than to die an old maid!"

The avowed purpose of the movement, however, is not the convenience nor the happiness of the "war brides," but a plan to restore the population, certain to be enormously depleted by the war. "Give us children or we perish; this is the tragedy of national existence," at least in times of war.

2nd Lieutenant E.F. Boyd

The only son of his parents, Edward Fenwick Boyd attended University College, Oxford, where he won a Football Blue, and went to France with the First Expeditionary Force in August 1914. He took part in the battles of the Mons, the Marne, and the Aisne, before falling in action on December 20, 1914, at the age of 24. He left no issue.

"They may not be happy children," says the *Chicago Herald*. "Many will never see their fathers. The mothers of many will die to give them birth, weak with suspense and fear and want. But they will be children! Say what you will of national honor, patriotism, all the rest! The supreme necessity of a nation is children.

"Breed before you die! It is our future that makes up those battalions and regiments of eager men so soon to know the freezing trench, the death rattle, and all the horrors of war. Leave us our future ere ye go."

The Sisters of the Fallen Who Never Became Mothers

After any war, and as a result of modern warfare, the number of women must necessarily exceed the number of men, the ratio of birth in the two sexes being always approximately equal, while the waste of men (except in desolated districts) is always greater. This disparity leads to drudgery as the lot of a greatly increased body of women. Furthermore, it is often the sisters of those who have died in combat who fail to wed, since the death toll has fallen on the better endowed men as a class, and the unmarried female relatives they leave behind will experience proportionately greater difficulty in finding a comparable mate from amongst those who survived the holocaust precisely because the latter tended to belong to a different and inferior segment of the population. Thus, contrary to popular belief, although the sisters of the fallen carry the same genes, they too are often condemned to be childless, and the nation's loss is thereby doubled.

(1) *The Independent.*
(2) *War Brides,* a drama by Marion Craig Westworth.

VI. WAR SELECTION IN THE ANCIENT WORLD

The Fall of Rome

"*The human harvest was bad!*" Thus the historian sums up the conditions in Rome in the days of the good emperor, Marcus Aurelius. By this he meant that while population and wealth were increasing, manhood had failed. There were men enough in the streets, men enough in the camps, menial laborers enough and idlers enough, but of good soldiers there were too few. For the business of the state, which in those days was mainly war, its men were inadequate.

In recognition of this condition we touch again the overshadowing fact in the history of Europe, the effect of "military selection" on the human breed.

In rapid survey of the evidence brought from history one must paint the picture, such as it is, with a broad brush, not attempting to treat exceptions and qualifications. The evil effects of military selection and its associated influences have long been recognized in theory by certain students of Social Evolution. But the ideas derived from the sane application of our knowledge of Darwinism to history are even now just beginning to penetrate the current literature of war and peace. In public affairs most nations have followed the principle of opportunism, "striking while the iron is hot," without regard to future results, whether of financial exhaustion or of race impoverishment.

The recorded history of Rome begins with small and vigorous tribes inhabiting the flanks of the Apennines and the valleys down to the sea, and blending together to form the Roman republic. They were men of courage and men of action, virile, austere, severe and dominant.[1] For this reason, Rome was long a republic. Free-born men control their own destinies. "The fault," says Cassius, "is not in our stars, but in ourselves that we are underlings." Thus in freedom, when Rome was small, without glory, without riches, without colonies and without slaves, she laid the foundations of greatness.

But little by little the spirit of freedom gave way to that of domination. Conscious of power, men sought to exercise it, not on themselves but on one another. Little by little this meant

Lieutenant M.C.N. Herbert

The only son of his parents, Malcolm Herbert was commissioned into the Gloucestershire Regiment in February 1914, went to France in August 1914, and was mortally wounded in the advanced trenches at Festubert, on December 22, 1914. He died, aged 20, without heirs.

aggression, suppression, plunder, struggle, glory and all that goes with the pomp and circumstance of war. So the individuality in the mass was lost in the aggrandizement of the few. Independence was swallowed up in ambition and patriotism came to have a new meaning, being transferred from hearth and home to the camp and the army.

In the subsequent history of Rome, we have now to consider only a single factor, the "reversal of selection." In Rome's conquests, *Vir*, the real man, went forth to battle and foreign invasion; *Homo*, the human being, remained on the farm and in the workshop and begat the new generations. "*Vir* gave place to *Homo*," says the Latin author. Men of good stock were replaced by the sons of slaves and camp-followers, the riff-raff of those the army sucked in but could not use.

The Fall of Rome was due not to luxury, effeminacy or corruption, not to Nero's or Caligula's wickedness, not to the futility of Constantine's descendants. It began at Philippi, where the spirit of domination overcame the spirit of freedom. It was forecast still earlier in the rise of consuls and triumvirs incident to the thinning out of the sturdy and self-sufficient strains who brooked no arbitrary rule. While the best men were falling in war, civil or foreign or remained behind in far-away colonies, the stock at home went on repeating its weakling parentage. A condition significant in Roman history is marked by the gradual swelling of the mob, with the rise in authority of the Emperor who was the mob's exponent. Increase of arbitrary power went with the growing weakness of the Romans themselves. Always the "Emperor" serves as a sort of historical barometer by which to measure the abasement of the people. The concentrated power of Julius Cæsar, resting on his own tremendous personality, showed that the days of Cincinnatus and of Junius Brutus were past. The strength of Augustus rested likewise in personality. The rising authority of later emperors had its roots in the ineffectiveness of the mob, until it came to pass that "the little finger of Constantine was thicker than the loins of Augustus." This was due not to Constantine's force, but to the continued reversal of selection among the people over whom he ruled. The Emperor, no longer the strong man holding in check all lesser men and organizations, became the creature of the mob; and "the mob, intoxicated with its own work, worshiped him as divine." Doubtless the last emperor, Augustulus Romulus, before the Goths threw him into the scrap-heap of history, was regarded by the mob and himself as the most god-like of the whole succession.

The Romans of the Republic might perhaps have made a history very different. Had they held aloof from world-conquering schemes Rome might have remained a republic, enduring even down to our day. The seeds of Rome's fall lay not in race nor in form of government, nor in wealth nor in senility, but in the influences by which the best men were cut off from parenthood, leaving its own weaker strains and strains of lower races to be fathers of coming generations.

"The Roman Empire," says Professor Seeley, "perished for want of men." Even Julius Cæsar notes the dire scarcity of men (δεινήν ὀλιγανθροπίαν), while at the same time there were people enough. The population steadily grew; Rome was filling up like an overflowing marsh. Men of a certain type were plenty, but self-reliant farmers, "the hardy dwellers on the flanks of the Apennines," men of the early Roman days, these were fast going, and with the change in type of population came the turn in Roman history.

"The mainspring of the Roman army for centuries had been the patient strength and courage, capacity for enduring hardships, instinctive submission to military discipline of the population that lined the Appennines."

"The effect of the wars was that the ranks of the smaller farmers were decimated, while the number of slaves who did not serve in the army multiplied," says Professor Bury. Thus "*Vir gave place to Homo*," thus the mob filled Rome and the mob-hero rose to the imperial throne. No wonder that Constantine seemed greater than Augustus. No wonder that "if Tiberius chastised his subjects with whips, Valentinian chastised them with scorpions."[2]

With Marcus Aurelius and the Antonines came a "period of sterility and barrenness in human beings." Bounties were offered for marriage. Penalties were devised against race-suicide. "Marriage," says Metellus, "is a duty which, however painful, every citizen ought manfully to discharge." Wars were conducted in the face of a declining birth-rate, and the decline in quality and quantity in the human breed engaged very early the attention of Roman statesmen. Deficiencies of numbers were made up by immigration, willing or enforced. Failure in

Lieutenant P.G. Chaworth-Muster

A professional soldier who was commissioned into the King's Royal Rifle Corps and served in Egypt, Chaworth-Muster arrived on the Western Front on August 12, 1914, and was wounded on September 1 at Villers Cotterets, near Mons. Recovering, he rejoined his battalion in time to be mortally wounded in January 1915. He died, aged 27, without issue.

quality was beyond remedy.

Says Professor Zumpt: "Government having assumed godhead took at the same time the appurtenances of it. Officials multiplied. Subjects lost their rights. Abject fear paralyzed the people and those that ruled were intoxicated with insolence and cruelty." "The Emperor possessed in the army an overwhelming force over which citizens had no influence, which was totally deaf to reason or eloquence, which had no patriotism because it had no country, which had no humanity because it had no domestic ties." "There runs through Roman literature a brigand's and barbarian's contempt for honest industry." "Roman civilization was not a creative kind, it was military, that is, destructive."

What was the end of it all? The nation bred Romans no more. To cultivate the Roman fields "*whole tribes were borrowed.*" The man with quick eye and strong arm gave place to the slave, the scullion, the pariah, whose lot is fixed because in him there lies no power to alter it. So at last the Roman world, devoid of power to resist, was overwhelmed by the Ostrogoths. "The barbarian settled and peopled the empire rather than conquered it. It was the weakness of war-worn Rome that gave the Germanic races their first opportunity." "A nation is like a bee," wisely observes Bernard Shaw, "as it stings it dies."

Seeck's Interpretation

In his monumental history of the "Downfall of the Ancient World" (*Der Untergang der Antiken Welt*), Dr. Otto Seeck of the University of Münster in Westphalia treats in detail the causes of such decline. He first calls attention to the intellectual stagnation which came over the Roman Empire about the beginning of the Christian Era. This manifested itself in all fields of intellectual activity. No new idea of any importance was advanced in science nor in technical and political studies. In the realm of Literature and Art also one finds a complete lack of originality and a tendency to imitate older models. All this, Seeck asserts was brought about by the continuous "rooting out ("*Austrottung*") of the best"[3] through war.

Such extermination which took place in Greece as well as in Rome, was due to persistent internal conflicts, the constant murderous struggle going on between political parties, in which, in rapid succession, first one and then the other was victorious. The custom of the victors being to kill and banish the leaders and all prominent men in the defeated party, often destroying their children as well, it is evident that in time every strain distinguished for moral courage, initiative or intellectual strength was exterminated. By such a systematic killing off of men of initiative and brains, the intellectual level of a nation must necessarily be lowered more and more.[4] In Rome as in Greece, observes Seeck: "A wealth of force of spirit went down in the suicidal wars." "In Rome, Marius and Cinna slew the aristocrats by hundreds and thousands. Sulla destroyed the democrates, and not less thoroughly. Whatever of strong blood survived fell as an offering to the proscription of the Triumvirate." "The Romans had less of spontaneous force to lose than the Greeks. Thus desolation came sooner to them. Whoever was bold enough to rise politically in Rome was almost without exception thrown to the ground. *Only cowards remained, and from their blood came forward the new generations.*[5] Cowardice showed itself in lack of originality and in slavish following of masters and traditions."

Certain authors, following Varro, have maintained that Rome died a "natural death," the normal result of old age. It is mere fancy to suppose that nations have their birth, their maturity and their decline under an inexorable law like that which determines the life history of the individual. A nation is a body of living men. It may be broken up if wrongly led or attacked by a superior force. When its proportion of men of initiative or character is reduced its future will necessarily be a resultant of the forces that are left.

Dr. Seeck speaks with special scorn of the idea that Rome died of "old age." He also repudiates the theory that her fall was due to the corruption of luxury, neglect of military tactics or over-diffusion of culture.

"It is inconceivable that the mass of Romans suffered from over-culture.[6] In condemning the sinful luxury of wealthy Romans we forget that the trade-lords of the fifteenth and sixteenth centuries were scarcely inferior in this regard to Lucullus and Apicius, their waste and luxury not constituting the slightest check to the advance of the nations to which these men belonged. The people who live in luxury in Rome were scattered more thinly than in any modern state of Europe. The masses

Second Lieutenant G.E. Burdekin

Articled as a solicitor, Geoffrey Burdekin volunteered his services with the outbreak of World War I and was commissioned into the Sherwood Foresters. He was killed by shell fire at Beuvry on January 26, 1915, at the age of 22, leaving no children.

lived at all times more poorly and frugally because they could do nothing else. Can we conceive that a war-force of untold millions of people is rendered effeminate by the luxury of a few hundreds?...Too long have historians looked on the rich and noble as marking the fate of the world. Half the Roman Empire was made up of rough barbarians untouched by Greek or Roman culture..."

"The Niobe of Nations, there she stands,
Crownless and childless in her voiceless woe;
An empty urn within her withered hands,
Whose sacred dust was scattered long ago!"
Byron.

Greece

Once Greece led the world in intellectual pursuits, in art, in poetry, in philosophy. A large and vital part of European culture is rooted directly in the language and thought of Athens. The most beautiful edifice in the world was the Peace Palace of the Parthenon, erected by Pericles, to celebrate the end of Greece's suicidal wars. This endured 2,187 years, to be wrecked at last (1687) in Turkish hands by the Christian bombs of the Venetian Republic.

But the glory of Greece had passed away long before the fall of the Parthenon. Its cause was the one cause of all such downfalls – the extinction of strong men by war. At the best, the civilization of Greece was built on slavery, one freeman to ten slaves. And when the freemen were destroyed, the slaves, an original Mediterranean stock, overspread the territory of Hellas along with the Bulgarians, Albanians and Vlachs, barbarians crowding down from the north.

The Grecian language still lives, the tongue of a spirited and rising modern people. But the Greeks of the classic period – the Hellenes of literature, art and philosophy – will never be known again. "Most of the old Greek race," says Mr. W.H. Ireland, "has been swept away, and the country is now inhabited by persons of Slavonic descent. Indeed, there is a strong ground for the statement that there was more of the old heroic blood of Hellas in the Turkish army of Edhem Pasha than in the soldiers of King George." The modern Greek has been called a "Byzantinized Slav." King George himself and Constantine his son are only aliens placed on the Grecian throne to suit the convenience of outer powers, being in fact descendants of tribes which to the ancient Greeks were merely barbarians.

It is maintained that "the modern Greeks are in the main descendants of the population that inhabited Greece in the earlier centuries of Byzantine rule. Owing to the operation of various causes, historical, social and economic, that population was composed of many heterogeneous elements and represented in very limited degree the race which repulsed the Persians and built the Parthenon. The internecine conflicts of the Greek communities, wars with foreign powers, and the deadly struggles of factions in the various cities had to a large extent obliterated the old race of free citizens by the beginning of the Roman period. The extermination of the Platæans by the Spartans and of the Melians by the Athenians during the Peloponnesian War, the proscription of the Athenian citizens after the war, the massacre of the Corcyrean oligarchs by the democratic party, the slaughter of the Thebans by Alexander and of the Corinthians by Mummius are among the more familar instances of the catastrophes which overtook the civic element in the Greek cities. The void can only have been filled from the ranks of the metics or resident aliens and of the descendants of the far more numerous slave population. In the classic period four-fifths of the population of Attica were slaves; of the remainder, half were metics. In 100 A.D. only three thousand free arm-bearing men were in Greece." (James D. Bourchier.)

"The constant little struggles of the Greeks among themselves made no great showing as to numbers compared to other wars, but they wiped out the most valuable people, the best blood, the most promising heredity on Earth. This cost the world more than the milling of millions of barbarians. In two centuries there were born under the shadow of the Parthenon more men of genius than the Roman Empire had in its whole existence. Yet this empire included all the civilized world, even Greece herself." (La Pouge.)

The downfall of Greece,[7] like that of Rome, has been ascribed by Schultz to the crossing of the Greeks with the barbaric races which flocked into Hellas from every side. These resident aliens, or metics, steadily increased in number as the free Greeks disappeared. Selected slaves or helots were then

Second Lieutenant B.H. Francis

The only surviving son of his parents, Lieutenant Francis left University College, Oxford, to volunteer for service and was commissioned into the Royal Scots Regiment later that year. He died in the trenches near La Basée on February 4, 1915, aged 19, leaving no issue. As a soldier wrote in a letter home: "we lost our Company Lieutenant last week . . . he was an awful nice man. He was a true Tommy Atkin's friend. He was too good to live, and too brave to lose."

made free in order to furnish fighting men, and again as these fell their places were taken by immigrants.

It is doubtless true at this day that "no race inhabits Greece," and the main difference between Greeks and other Balkan peoples is that, inhabiting the mountains and valleys of Hellas they speak in dialects fo the ancient tongue. Environment, except through selection and segregation, cannot alter race inheritance and the modern "Greeks" have not been changed by it. In other words, it is quite true that the Greece of Pericles owed its strength to Greek blood, not to Hellenic scenery. When all the good Greek blood was spent in suicidal war, only slaves and foreign-born were left. "'Tis Greece, but living Greece no more."[8]

Furthermore we do not know that even the first Hellenes of Mycenæ were an unmixed race, or that any unmixed races ever rose to such prominence as to command the world's attention. We do know that when war depletes a nation slaves and foreigners come in to fill the vacuum, and that the decline of a great race in history has always been accompanied by a debasing of its blood.

Yet out of this decadence natural selection may in time bring forward better strains, and with normal conditions of security and peace, nature may begin again her work of recuperation.

In the fall of Greece we have another count against war, scarcely realized until the facts of Louvain and Malines, of Rheims and Ypres have brought it again so vividly before us. War respects nothing, while the human soul increasingly demands veneration for its own noble and beautiful achievements. As I write this, there rise before me the paintings in the "Neue Pinakothek" at Münich, representing the twenty-one cities of Ancient Greece, from Sparta to Salamis, from Eleusis to Corinth, not as they were, "in the glory which was Greece," not as they are now, largely fishing hamlets by the blue Ægean Sea, but as ruined arches and broken columns half hid in the ashes of war, wars which blotted out Greece from world history.

"Such are the sights, the sorrows fell,
About our hearth – and worse, whereof I may not tell.
But all the wide town o'er,
Each home that sent its master far away
From Hellas' shore,

Feels the keen thrill of heart, the pang of loss, to-day.
  For, truth to say,
The touch of bitter death is manifold.
Familiar was each face, and dear as life,
  That went into the war,
But thither, whence a warrior went of old,
  Doth naught return –
Only a spear and sword, and ashes in an urn.
  For Ares, lord of strife,
Who doth the swaying scales of battle hold,
War's money-changer, giving dust for gold,
  Sends back, to hearts that held them dear,
Scant ashes of warriors, wept with many a tear,
Light to the hand but heavy to the soul;
    Yea, fills the light urn full
  With what survived the flame –
Death's dusty measure of a hero's fame."[9]

(1) *Virilis, austerus, severus, dominus,* good old words applied by Romans to themselves.

(2) The point of this is that the cruel Tiberius was less severe on the Romans of his day than was the relatively benevolent Valentinian on his decadent people.

(3) "Die Ausrottung der Besten, die jenen schwächeren Volken die Vernichtung brachte, hat die starken Germanen erst befähigt, auf den Trummern der antiken Welt neue dauerende Gemeinschaften zu errichten." (Seeck.)

(4) The history of Korea reveals much the same condition. Three hundred years ago this country had reached a considerable degree of civilization. Its conquest by Hideyoshi, Shogun of Japan was followed by a vigorous reaction, in which the Japanese armies were flung out of Korea and the Japanese fleet destroyed. At that time in art matters at least, the Koreans were more advanced than the Japanese. The Buddhist temples and the palaces of Kyoto and Nagoya are modeled after similar buildings in Seoul, being, in fact mostly built by Korean artisans.

In modern times, until the country was taken over by Japan, the government of Korea was singularly inert and correspondingly cruel, while the people though individually fairly intelligent had come to lose all initiative due to the persistent practice of the rulers of beheading all persons opposed to their policies. Similar customs widely spread in earlier times in Europe as well as in Asia, must have been a large factor in the extirpation of initiative. The old English habit of sending "to the Tower" those lords or ministers the crown found troubesome was another form of the same costly waste of ability from which in ruder times no nation was free.

(5) *Author's italics.*

(6) "Damitsprechend hat man das Wort 'Ueberkulture' überhaupt erfunden, als wenn ein zu grosses Maass von Kulture überhaupt denkbar wäre."

(7) Certain recent writers who find in environment the causes of the rise and fall of nations, ascribe the failure of Greece to the introduction in Athens and Sparta of the malaria-bearing mosquito. As to the facts in question, we have little evidence. But

2nd Lieutenant W.H. Clarke

The second son of a regular officer, Sir Edward Clarke, of Rossmore, County Cork, Lieutenant Clarke went to France with the Worcestershire Regiment in December 1914, and died in Flanders under heavy fire three months later, on March 12, 1915 – all the officers in his Company being killed on that day. He died without issue, and his elder brother was also killed in the war.

while the prevalence of malaria may have affected the general activity of the people, it could in no way have obliterated the mental leadership which made the strength of classic Hellas, nor could it have injected its poison into the stream of Greek heredity.

(8) In contrasting a new race with the old – as the modern Greeks with the incomparable Hellenes – we must not be unjust to the men to today whose limitations are evident, contrasted with a race we know mainly by its finest examples. In spite of poverty, touchiness and vanity characteristic of the modern Greek, there is good stuff in him. He is frank, hopeful, enthusiastic. The mountain Greek, at least, knows the value of freedom, and has more than once put up a brave fight for it.

(9) From the chorus to *Clytemnestra* in the *Agamemnon* of Æschylus. (Translated by E.D.A. Morshead.)

VII. WAR AND WAR SELECTION IN WESTERN EUROPE

France

Europe had no finer human stock than that of France, and no modern people has suffered more from the ravages of war and glory. The Gauls as they appear in early history were a Celtic race. Conquest made them Gallo-Roman. Later, especially in the North and East, their blood was strengthened by Teutonic strains, – the Normans from Scandinavia and the Franks from Central Germany. In later days a large influx from Germanic Alsace has made German names common in French society.

Certain effects of war on the French people have been already considered in a previous chapter. Through reversal of selection by war France lost men of initiative. But nations also possess powers of recuperation, and regenerative processes have been evident in France for the last twenty years. Peace and security, industry and economy enable the natural forces of race regeneration to operate. The nation had been sorely wounded by her own sons. She has been making a healthy recovery.[1]

In the Wiertz gallery in Brussels is a striking painting dating from the time of Napoleon, called "A Scene in Hell" (*Une Scène dans l'Enfer*). It represents the great marshal with folded arms and face unmoved descending slowly to the land of the shades. Before him filling all the background of the picture, their faces expressing every form of reproach, are the men sent to death before their time by his unbridled ambition. Four millions there were in all, more than half of them French. And behind the legions shown or hinted at, one seems to discern the millions on millions who might have been and are not – the huge widening wedge of the possible descendants of those who fell in battle, youth without blemish (*"L'élite de l'Europe"*), made "flesh for the cannon" in the rush of Napoleon's great campaigns.

They came from the farm, the workshop, the school, men from eighteen to thirty-five years of age at first, but afterwards the older and the younger. "A boy will stop a bullet as well as a man," said Napoleon. "The more vigorous and well-born a young man is," said Professor Haeckel, "the more normally con-

Captain E.J.F. Johnston

A professional soldier who was commissioned into the Royal Scots Regiment and served in the South African War, receiving two Medals and five Clasps. He went to the Western Front on the outbreak of war, and was described as "far and away the best loved officer I have met" and "the men put complete trust and confidence in him. At night, instead of sleeping, he would crawl along to the men at the listening posts to see how they were getting on." Captain Johnston was killed on April 12, 1915, aged 32. He left a widow but no children.

stituted, the greater his chance to be slain by musket or magazine, the rifled cannon and other similar engines of civilization." "Napoleon," says Seeck, "in a series of years seized all the youth of high stature and left them scattered over many battlefields, so that the French people who followed them are mostly of smaller stature. More than once since Napoleon's time has the military limit been lowered."

In the career of Napoleon campaign followed campaign, against enemies, against neutrals, against friends. Conscription followed victory, both victory and conscription debasing the human species. Again conscription after conscription. "Let them die with arms in their hands. Their death is glorious, and it will be avenged. You can always fill the places of soldiers." "A great soldier like me doesn't care a tinker's damn for the lives of a million men." Still more conscription! After Wagram, France began to feel her weakness, the "Grand Army" being no longer the army which had fought at Ulm and Jena. "Raw conscripts raised before their time and hurriedly drafted into the line had impaired its steadiness."[2]

After Moscow, homeward "amidst everdeepening misery they struggled on, until of the six hundred thousand men who had proudly crossed the Niemen for the conquest of Russia, only twenty thousand famished, frost-bitten, unarmed specters staggered across the bridge of Korno in the middle of December." "Despite the loss of the most splendid army marshalled by man, Napoleon abated no whit of his resolve to dominate Germany and discipline Russia....He strained every effort to call the youth of the empire to arms...and 350,000 conscripts were promised by the Senate. The mighty swirl of the Moscow campaign sucked 150,000 lads of under twenty years of age into the devouring vortex." "The peasantry gave up their sons as food for cannon." But "many were appalled at the frightful drain on the nation's strength." "In less than half a year after the loss of half a million men a new army nearly as numerous was marshalled under the imperial eagles. But the majority were young, untrained troops, and it was remarked that the conscripts born in the year of Terror had not the stamina of the earlier levies. Brave they were, superbly brave, and the emperor sought by every means to breathe into them his indomitable spirit." "Truly the emperor could make boys heroes, but he could never repair the losses of 1812." "Soldiers were wanting, youths were dragged forth." "To fill hell with heroes," – in these words some one has summed up the life-work of Napoleon. "*J'ai cent mille hommes de rente*," "my income is a hundred thousand men," said Napoleon. But to a terrible degree he lived beyond his income.

French writers have been frank in the discussion of national deficiencies and mistakes. They have wished to conceal nothing from France and therefore nothing from the world. Their admissions have been exaggerated by unfriendly critics. It has been claimed that modern France, with the other Latin nations, is a "decadent state."[3] that she has passed her prime and is now in the weakness and sterility of old age, her place as the dominating force on the continent of Europe having been yielded to a younger and more aggessive power. Decrepitude in a nation is due not to age, but to the operations of war, as we have several times insisted, followed by the loss of its best strains of blood and their replacement by recruiting from immigrants of the weaker races.

Certain features of French life have been indicated as evidences of injury from reversal of selection. The birth rate in France already low has been steadily falling. This is apparently a result of the survival of the cautious, for Napoleon's dashing grenadiers could hardly be imagined to limit their families for prudential reasons

Edmond Demolins in a clever books asks: "In what constitutes the superiority of the Anglo-Saxon?" He finds his answer in the false standards of French life, in defects of training and of civic and personal ideals. The desire for seats in a government bureau and for similar safe places of routine and without initiative has been termed in Italy "*Impiegomania*," – the "craze for sitting down." The eagerness to secure such positions is said to be a besetting sin of the youth of both Italy and France. But the fault may be due to over-centralization of government, too many officials and too little opportunity in the provincial centers, rather than to any fault in the nature of the individual man. Nationalization of effort, whether through socialism or through "efficient organization," must contribute to the spread of "impiegomania."

If the strictures of Demolins be true in any degree, this may be the interpretation. Inferior standards are the work of inferior men. Great men there are in France, and these have persistently

Captain M.A. Fitzroy

Leaving his studies at Oriel College, Oxford, Captain Fitzroy was commissioned into the Seathforth Highland Regiment in September 1914, and took part in the Battle of Neuve Chapelle where he was lightly wounded. He was killed on April 16, 1915 while manning an advanced listening post. Aged 19, he left no children.

turned the nation's face toward the light since Demolins' book was written. War's effect has been to rob her of her due propro-tion of leaders, but not to dilute or to weaken the message of those who survive. The evolution of a race is always selective, never collective. Collective evolution among men or beasts, the movement upward or downward of the whole as a whole, irre-spective of training or selection, is never a fact. As La Pouge has said: "It exists in rhetoric, not in truth nor in history."

Spain

The Spain of today is not the Spain of 1493 to whom the Pope assigned half the seas of the world. Old Spain drooped long ago, exhausted with intolerance, sea power and empire. Now that modern Spain has been deprived of the last vestige of Imperial control, she is slowly recuperating on a foundation of industry and economy.

In 1630, the Augustinian friar, La Puente, thus wrote of the fate of Spain; "Against the credit for redeemed souls I set the cost of armadas and the sacrifice of soldiers and friars sent to the Philippines. And this I count the chief loss; for mines give silver, and forests give timber, but only Spain gives Spaniards, and she may give so many that she will be left desolate, and constrained to bring up strangers' children instead of her own." "This is Castile," said a Spanish knight; "she makes men and wastes them." "This sublime and terrible phrase sums up Span-ish history."[4]

"Everything has happened that could happen," says Havelock Ellis, "to kill out the virile, militant, independent elements of Spanish manhood. War alone, if sufficiently prolonged and severe, suffices to deplete the nation of its most vigorous stocks. 'The warlike nation of today...is the decadent nation of tomor-row.' The martial ardor and success of the Spaniards lasted for more than a thousand years. It was only at very great cost that the Romans subdued the Iberians and down to the sixteenth century, the Spaniards were great soldiers. The struggle in the Netherlands wasted their energies and then finally at Rocroy, in the middle of the seventeenth century, the Spanish infantry that had been counted the finest in Europe went down before the French, and the military spendor of Spain vanished" (*The Soul of Spain*).

It is a question whether Spain suffered most from the scattering of her strong men overseas, or from her perpetual struggles in Europe.

Germany

Germany suffered perhaps scarcely less than France from the wars of Louis XIV and of the two Napoleons. German writers, however, have been much less frank than the French and also less lucid in discussing their national disabilities. They have given but scanty records of the racial waste their wars have involved. Moreover, the organization of modern Germany, a socialist state under military domination, has tended to minimize the visible distinctions among racial strains. Every man has his place. It is not easy to fall below one's class, correspondingly difficult to rise. Universal compulsory education, technical as well as academic, saves even the feeble from absolute incompetence.

The victory over France in 1871 has had the effect of intensifying the military spirit of Germany, and of making its extension appear an integral part of the nation's commercial and industrial growth. This fact operates toward final disaster, for whether successful or not in the struggle with the allied powers, the aggregate result will be of the nature of terrible defeat. When the record is summed up it may appear that Germany rather than France is the final sufferer from the Franco-Prussian war and the "blood and iron" policy of Bismarck and his successors.

That the present war will cut deep into the best stock of Germany no one can doubt. As to the effects of their great wars of the past, the most important studies known to us are those of Dr. Seeck and Dr. Karl G. Rendtorff. Professor Rendtorff[5] furnishes for our purpose the results of his researches in the decline of German literature following the wars of the thirteenth century. In brief, he shows that the period from 1170 to 1230 marks a climax in early German culture. The thirteenth, "the greatest of centuries," saw the culmination of early poetry, of the epic of chivalry, of military prowess, of imperial greatness.

The period came to a sudden end with the downfall of Knighthood and the consequent cessation of intellectual life. The Knights were the pick of the nation. Knighthood was, in

Private G.S. Grundy

The only son of his parents, Geoffrey Grundy was articled to a firm of solicitors when war broke out, but volunteered his services promptly, and like many city professionals joined the historic Honorable Artillery Company on August 6, 1914. He sailed for France the next month, moved up to the Front in November, and was killed in action on April 14, 1915. A comrade wrote: "he was a great favourite with the regiment: there was no crouching below the parapet when on guard: he stood up square and kept his watch conscientiously, in spite of the bullets." He died aged 28, without issue.

some degree, a democratic institution open to any one, even a serf who distinguished himself by physical courage or by mental power. Thousands of Knights fell in the bloody wars waged by the Hohenstaufens in Italy, slain in battle, yielding to disease and poison. Thousands more died in the Crusades, through epidemics in the ports and battles in Palestine. Border warfare existed everywhere, taking its toll while many of the finest minds escaped to the shelter of the Church. In all these episodes, a "survival of the fittest" was impossible. The inevitable result of the destruction of the intellectual and physical leadership of the nation was the mental decay of which the most glaring evidence is found in the contemporaneous decline of German poetry. Later, in the rise of the Burghers, appeared new life, a literature without imagination or originality, but to be later followed by the great creative work of architects, sculptors and painters.

* * *

It is apparent that organized warfare, properly a conflict of soldier with soldier, has the final effect of minimizing the very qualities it demands. A massacre differs from war in that its operations are not confined to soldiers. It involves old and young, sparing neither age nor sex. Only those escape who are vigorous enough to run away and to endure starvation and exposure. In the Thirty Years' War of the seventeenth century, a continuous massacre, the population of Germany was reduced from about sixteen millions to six millions.

It is claimed by some writers, notably by Seeck and Ammon, that the slaughter, however terrible from the standpoint of humanity, caused no injury to the quality of the German race. Seeck insists that this is indeed the fact. According to him, in massacres, more weak than strong are killed and consequently, while the numbers are greatly reduced, the average physical and intellectual strength of the nation is thereby rather enhanced:[6]

To quote from Seeck: "A systematic extermination of exceptional people, such as we find was carried out in Greece and Rome, must produce a race of cowardly and mediocre men, on the other hand, when a terrible war sweeps over a nation, indiscriminately killing thousands upon thousands, we may expect the opposite result....For if out of every one hundred thousand *strong men* eighty thousand were killed, surely out of every one hundred thousand weeklings, at least ninety thousand or perhaps ninety-five thousand were killed." He cites a number of examples to prove his theory and claims that usually a century or so after a calamity such as this, the nation bears its finest intellectual fruit. He points to Spain which, during a period of peace lasting for more than fifteen hundred years, furnished nothing valuable in art or science, but which, at the time of Cervantes and Velasquez, produced a generation of heroes, exactly a hundred years after a succession of terrible civil wars and massacres had swept over the country. Germany lost three-fourths of her population during the Thirty Years' War, the Netherlands suffered a similar fate during the Wars of Liberation, England during the Wars of the Roses, and France during the War of the Huguenots. But a century after the worst slaughters, Goethe and Kant were born in Germany, Shakespeare and Bacon in England, Molière and Bayle in France, while Grotius and Rembrandt appeared in Holland even before a century had passed. As further proof that a great civil catastrophe acts beneficially, he cites Northern Italy, which towards the close of the Middle Ages was again and again the scene of massacres which left Southern Italy untouched; and yet it was Northern Italy which gave the Renaissance to the world while the southern districts have hardly done anything at all for the glory of the nation.[7] Finally he asserts that among the ancient Germans when individual tendencies were too strong to permit the founding of a state, the extermination of the best, that is of the strongest personalities, really was necessary for the establishment of any government able to hold its own against its own citizens.

Even if correct, the facts would not extenuate massacre, the most revolting and debasing of all forms of war, and it is incredible that the birth of great men like Shakespeare and Goethe is dependent on the killing of a multitude of weaker ones. Moreover, the details of such history are very imperfectly known. We have little that can be spoken of as statistics. Christopherus Grimmelshausen's story of "*Simplicius Simplicissimus*," the most elaborate of these records, is one long tale of debauchery and horror. The ancient "Chronicles of Thann" relate something of how upper Alsace was ravaged. There towns were destroyed never to be rebuilt. In one commune, there was not for twelve years a wedding, nor for fifteen a baptism. "So often as the

Captain A.H.P. Cruickshank

Commissioned from the Royal Military Academy, Woolwich, in 1901, Cruickshank first served with the Royal Garrison Artillery before serving in the British Indian Army with the 32 Sikh Pioneers. Despatched to France at the outbreak of the war, he was wounded on November 23rd 1914 at Givenchy and again, mortally, on April 27, 1915 near Ypres. Captain Cruickshank, who was married in July 1914, left a widow but no children. He was 31 years old at his death.

Swedes gave battle to the Imperialists, so often did the Imperialists make war on the Swedes. It was an endless massacre."

"In the period of savagery," observes Novicow, "war between tribes was without pity. The victors destroyed the conquered to the last man, then married the women. To a certain degree, the result was favorable to the race, but on this condition that no one among the conquerors was killed. This evidently was never the case, hence, after any battle, the number of fine men who might have wives was always reduced. Here again war produces a reversal of selection....The pretended benefit of war, even in savagery, disappears wholly with the advent of civilization."

(1) "Land, money, tradition and prestige," says Professor Albert Léon Guérard (*French Civiliation in the Nineteenth Century*, 1912), "would be naught if the people has lost its soul. Their wealth would pass into stronger hands, and their prestige to contempt. Once, about twenty years ago, the French themselves wondered if it had not come to that. The cry of a decadence was raised by malevolent rivals, by sensationalist, by "æsthetics" in quest of a new pose, by earnest patriots who had lost their star. When a belated echo of this reaches us now, how faint and strange and silly it sounds! For the one great asset of the French is their indomitable vitality. Even in wasteful conflict one cannot fail to admire the evidence of power. In the twentieth century as ever before the French are among the pioneers.

"I do not see France as a goddess, austere and remote. I see her intensely human, stained with indecencies and blasphemies, scarred with innumerable battles, often blinded and stumbling on the way, but fighting on undismayed, for ideals which she cannot always define. An old nation? A wounded nation? Perhaps, but her mighty heart is throbbing with unconquerable life."

(2) This quotation and those athat immediately follow are from the *History of Napoleon I*, by J.H. Rose.

(3) The French people "have sunk to so low a level in all the virtues of a strong and proud nation that, from the military standpoint, it must be regarded as a doubtful pleasure to have to fight such a people." (*Berliner Post*, April 21, 1913, quoted by W.H. Dawson.)

(4) Captain G.G. Calkins.

(5) See Appendix.

(6) Once after a great drought in Australia, a large proportion of the sheep died. Sheep owners naturally preserved the most valuable individuals and the average yield of wool in the next generation is said to have increased by one-third. This was an example of drastic selection, and something like this may have taken place in the Thirty Years' War.

(7) Seeck here seems to overlook the fact that the people of Naples and Sicily were of a different and much less virile stock than the Florentines and Lombards.

VIII. WAR SELECTION IN GREAT BRITAIN

Cost of Empire

Not long ago, in England, a parliamentary commission was set to inquire into the fact of "national deterioration." Of this there were various sorts of evidence. The yeomanry were disappearing. The slums of London, Manchester, Liverpool, Birmingham, and Sheffield, were centers of sweat-shops and child labor, of wasting over-work, of infant mortality, of malnutrition, of sodden drunkenness with helpless old age. And in the higher classes, one heard of "flannelled oafs" and heedless sportsmen to whom a cricket match was of more worth than the conservation of empire. Much of this was complacent self-criticism, a favorite amusement with the literary classes of England. Some of it had the political purpose of discrediting the government, but behind it all rests a certain neglected residuum of truth. For Great Britain has paid in full the costs of Empire.

In the Norse mythology, it was the Mitgard serpent which reached around the world, swallowed its own tail and held the earth together. England has been the Mitgard-serpent of history. She has made this in a sense a British world. Her youth have gone into all regions where free men can live. Everywhere they have built up free institutions held together by the British cement of coöperation and compromise. England carried her "Pax Britannica," with its semblance of order and decency, to barbarous lands, mixing with it enough of freedom to give permanence to her rule. She has made it possible for Englishmen to trade with savages. "What does he know of England, who only England knows?" For the activities of the Greater Britain of whom we of America form an integral part, outweigh those confined to the little island from which the British people set forth to inherit the Earth.

What has it all cost? For every great race sacrifice takes toll in race exhaustion. The loss will not appear in the decline in ability of the statesmen and scholars who remain. It will show itself in the relative fewness of strong men with a proportionate increase of weaklings and wastrels.

Much of the force of England has gone out to America and to her self-governing commonwealths, the forceful young democ-

Lieutenant C.C. Egerton

A professional soldier who was commissioned into the Duke of Wellington's West Riding Regiment in 1905, Egerton arrived in Flanders in August 1914. Attacking Hill 60 under heavy shell fire, he was killed on April 18, 1915, at the age of 30. Married in 1914, he left a widow but no children.

racies still proud of British traditions, even while escaping from the worst of these, the legalization of privilege. But a man is a man wherever he may live, and the occupation of Canada, Australia, New Zealand and South Africa has been a source of continued strength to the mother country.

With India the case is less certain. Men have asked, "What has Britain done for India"? We may admit that she has done much, and her work, improving with experience, grows more successful with the lapse of time. But one may rather ask: "What has India done for Britain"? The answer to this is not so clear and much loss as well as gain must enter into the calculation. India has enriched England – that small part of England which is engaged in over-seas trade. The men who have gained, like the Sassoons of the opium trade for example, are not as a rule those who share their fortunes with the people who have been taxed to make such fortunes possible. India has furnished employment for thousands of young Englishmen ("outdoor relief for sons of good families") and it has opened graves for thousands of British yeomen and British gentlemen, men of spirit whom Britain could ill afford to spare. A returning officer once said to me, "I have seen men who might have been makers of empire, die like flies in India."

Says Franklin: "The profits of no trade can even be equal to the expense of compelling it by force of armies." But the profits of the trade obtained through compulsion go to the undeserving few. The cost of compulsion in blood and in gold falls on the nation.

Can we measure the cost of British Imperialism?

"There's a widow in sleepy Chester
Who mourns for her only son;
There's a grave by the Pabeng River. –
A grave which the Burmans shun."

To know why Chester is sleepy, let us ask her red sandstone cathedral, "What of your dead"? The long array of bronze tablets tells the story. They bear the names of the fine-spirited yeomanry whose lives Cheshire has spent for empire. In almost every English cathedral and in almost every parish church in England and Scotland we read the same story. Britain has exchanged her young men in India for bronze tablets at home.

Says Alfred Noyes, speaking of England:

> "It is only my dead that I count,
> She said and she says to-day."

Here are names of sturdy farmers, of gentlemen's sons from Eton and Rugby, from Harrow and Winchester, of scholars from Oxford and Cambridge, all lost in some far-off war. Their bodies rest in India, in Burmah, in Afghanistan, in the Transvaal. "At home," they are remembered. What would have been the effect on England if all of these and their potential descendants could be numbered among her sons today?

Says Havelock Ellis: "The reckless Englishmen, who boldly sailed out from their little island to fight the Spanish Armada, were long since exterminated; and an admirably prudent and cautious race has been left alive."

> "We have fed our sea for a thousand years,
> And she calls us, still unfed;
> Though there's never a wave of all her waves,
> But marks our English dead.
>
> We've strewed our best to the waves' unrest,
> To the shark and the sheering gull.
> If blood be the price of Admiralty,
> Lord God, we ha' paid it in full!"

"We admit," says Prof. J. Arthur Thomson, "that wars have been necessary and righteous – especially necessary, and that they may be so still, but this opinion does not affect the fact that prolonged war in which a nation takes part is bound to impoverish the breed, since the character of the breed always depends on the men who are left. The only thing a nation dies of is lack of men and is there not disquieting evidence of the increase of incapables"?

The Picked Half-Million

"The picked half-million!" This phrase Mr. William T. Stead applied to the University men of Great Britain. "It is theirs to command and the world obeys." In the Great War now raging

Second Lieutenant J.C.D. Brown

Immediately following the outbreak of war, Lieutenant Brown was commissioned into the Durham Light Infantry, and went to the Front before even completing Divisional Training, where he was twice wounded while on advanced reconnaissance at the Battle of Ypres. He died from his wounds on April 27, 1915, aged 21, leaving no children.

Lieutenant T. Bailward

Another professional soldier who fell early in the war. Lt. Bailward was commissioned from Sandhurst in 1906, and served with the 26th King's Own Light Cavalry in the British Indian Army. In 1913 he was appointed A.D.C. to the British Governor of Madras, but was killed in Iraq on April 29, 1915 while serving with the 7th Hariana Lancers against the Turks. Aged 27 at his death, he left no children.

their blood is being wasted as never before in history. In Oxford (March 1915) not a "Blue" is left, not an athlete, only about eight hundred men out of more than four times that number for whom the authorities made ready last July. In an article, "*Oxford at War*," Professor L.P. Jacks, editor of *Hibbert Journal*, makes the following statements:[1]

"...It is probable that out of a normal total of 3,500, not more than 750 will be returning, certainly not more than 1,000. And of those who do return the number will be rapidly reduced as the pending commissions are allotted....The best – morally, intellectually, physically – have gone. With rare exceptions only the weaklings remain....

"The 'Roll of Honor' of Oxford men who have fallen in the war is already very long and it lengthens day by day. Oxford hardly dares to count its dead....I return to my work and presently another regiment passes, marching with a different step.... These, I think, must be raw recruits. But, no! This is the corps humorously known in the University as the 'Last Ditchers' – a body composed for the most part of professors and dons, some of them well advanced in life. There in the second rank is the Poet Laureate; there is Sir Walter Raleigh; there is Professor Gilbert Murray; there is Mr. Godley, the University orator, seemingly in command, and many others known to fame."

At Cambridge,[2] the same. Even the professors are engaged in drill or practicing with rifle or pistol. The Universities of Scotland, the municipal Universities of England, all tell the same story. The German Universities are marking time, and those of France scarcely even that. Young French instructors in American Universities have gone quietly back to the colors, and are on the firing line on the Aisne and the Marne.

Lloyd-George proudly asserts: "Our new army will be the most democratic and the most self-sacrificing that has ever rallied to a nation's colors....And fine soldiers, verily! All the pick of the nation, the best and the bravest of all classes of society, intellectuals as well as workmen, rich as well as poor, the élite of our trade unionists, as well as our most brilliant scholars of Oxford and Cambridge, the bench and bar as well as the shop, the factory as well as the club, have furnished these hundreds of thousands of vigorous men of from twenty-one to thirty-six years, with whom my colleague, Lord Kitchener, has formed his new army."[3]

One can hardly blame the Chancellor of the Exchequer for his point of view. His sole thought is of speedy victory, not of the nation's ultimate stability. Considering the long future of the nation only, not one scholar, not one skilled laborer ought be be sacrificed. In London, during the first months of the war, the writer saw troops going to the front and recruits to training camps, – strong, serious, well set-up Englishmen, reinforced by stronger and often more vigorous Scots. Left on the streets were many thousands of men war could not use. Many Londoners reserved to be fathers of a large part of the next generation are not well put together; the rejects form a strain from which England might well wish to be free. Nor is it so in London alone. The same conditions occur in a degree in every great city and in every war-swept land. "Father a weed, mother a weed, do you expect the daughter to be a saffron root?"

We still have but scanty details of recent battles, for the censors have drawn behind the line an impenetrable veil, and the events of current history are as obscure as those of the Middle Ages. Yet this is certain that the flower of our civilization is falling as never before, and the "gaps in our picked and chosen" will not be filled for a century. Will H. Irwin, of New York, a competent observer in the field, assures us that the long-drawn battle of Ypres cost England 50,000 out of 120,000 men engaged. The French and Belgian loss he estimates at 70,000 killed and wounded, that of the Germans at 375,000. "In that one long battle, Europe lost as many men as the North lost in the whole Civil War."

Scotland

In Scotland, the martial spirit has generally been stronger than in England. It is a trait of the emotional Celt of the Highlands as well as of the calculating Scotsman of the valley towns. The facts as to war-selection in Scotland have been vigorously set forth by Dr. James A. Macdonald.

"Scotland," says Macdonald, "speaks from long and sad experience. Every heathery hill looks down on a glen that, generation after generation, sent in answer to the fiery cross and pipes of war the best its home had bred. On those moors and through those intervals life at best was hard. The weaklings died in infancy. By the law of the survival of the fittest there

Captain J. Goold Adams

The only son of the Archdeacon of Derry in Ireland, Captain Adams was commissioned into the Leinster Regiment in 1903, and served in West Africa. Moving with his regiment to France in December 1914, he was killed by shell fire at the Battle of Hill 60, near Ypres, on May 4, 1915. Captain Adams left a widow but no children.

was bred a race of giants whose kilted regiments, every man six feet or more, were the pride of their race and the glory of British arms.

"In the awful days of the Forty-five, out of this very Glenurquhart eight hundred men of the clansmen's mold marched to Culloden for their 'Bonnie Prince Charlie.' But a fortnight ago among those who marched out to 'Leaving Glenurquhart,' not a corporal's guard, though they looked their best from Loch Ness to Corrimony, could pass the heroic standard of the olden days. Grants[4] from that glen and from Strathspey stained with their blood the marble palaces of India, and saved the honor of humanity in the awful days of the mutiny; but today few of their clan are left 'in their ain dear glen.' The sturdy Chisholms are gone from Strathglass. Wild and high, as through Belgium to Waterloo a hundred years ago, the 'Cameron's Gathering'[5] rose this very month when Lochiel called for his men, but how many had the 'biological' excellence of the clan 'what time the plaided chiefs came down to do battle with Montrose'? The Mackenzies today are few at Lochbroom. In the gloaming glens of the West Highlands there is a silence deep as death where once a thousand Campbell would start up in the night at the call of Argyll. No Lord of the Isles who sleeps in Iona could ever again gather a clan worthy his tartan, though he blew all night on the pibroch of Donald.

"Tell me, have the fittest survived? Go through their cities and over their moors and down their glens. More than 800 kilted soldiers of the giant mold went out of my ancestral glen at Culloden Moor; up and down that glen I have gone without seeing a corporal's guard of the olden type. In vain I looked for them even in Inverness itself.

"They went out, these Highland clans, wherever the Royal Standard flew. Again those Highland clans go out, the best and bravest of their breed, and they never come back. Biology does the rest. War's commercial dislocations and war's financial ruin are bad enough, but war's biological reaction is damage beyond repair.

"Is war a 'biological necessity'? Let Scotland answer. Never since the days of the Stuarts has Scotland, and especially the Scottish Highlanders, been free from the toll taken by the recruiting sergeants for Britain's army. The history of the Celts is in one sentence: 'Forever they went out to battle, and forever they fell.'"

(1) *Nation*, January 28, 1915.

(2) "The *Cambridge Review*, has published the names of the past and present members of the university who are serving in some capacity in the British army. The number amounts to 7,237, and they are distributed amongst the colleges as follows: Trinity College, 1,840; Pembroke, 760; Gonville and Gaius, 616; Clare, 535; Kings 436; Jesus, 385; Emanuel, 371; Christ's, 359; St. John's, 337; Trinity Hall, 328; Magdalene, 214; Queen's, 179; Sidney Sussex, 154; Peterhouse, 140; Downing, 126; Selwyn, 126; St. Catherine's 117, Corpus Christi, 109; Fitzwilliam Hall, 90; honorary graduates, 16." (*School and Society*, March, 1915.)

"Of 1,400 students in the École des Beaux Arts, 1,300 went out at the first call, and three-fourths of those who survive are in the trenches. They left behind only a few cripples and foreigners, such as Russians, natives of the Balkans, and occasional American and Englishmen." (Will H. Irwin.)

(3) From interview with H.B. Needham; *Collier's*, February 27, 1915.

(4) "When the good Lord was making Adam, even then the Clan Grant was as numerous as the heather on the hills." (Bailie Grant.)

(5) "Proudly they march, but each Cameron knows,
He may tread the heather no more."

Set in this stormy northern sea,
  Queen of these restless fields of tide,
England! *what shall men say of thee,*
  Before whose feet the worlds divide?

And thou whose wounds are never healed,
  Whose weary race is never won,
O Cromwell's England! must thou yield
  For every inch of ground a son?

What profit that our galleys ride,
  Pine-forest-like, on every main?
Ruin and wreck are at our side,
Grim warders of the House of Pain.

Where are the brave, the strong, the fleet?
  Where is our English chivalry?
Wild grasses are their burial-sheet,
  And sobbing waves their threnody.

Peace, peace! we wrong the noble dead
  To vex their solemn slumber so:
Though childless, and with thorn-crowned head,
  Up the steep road must England go.
Ave Imperatrix; *Oscar Wilde.*